Power Quality Monitoring, Analysis and Enhancement

Power Quality Monitoring, Analysis and Enhancement

Editor

Lucas Formann

scitus
academics

Power Quality Monitoring, Analysis and Enhancement
Edited by **Lucas Formann**

ISBN: 978-1-68117-613-0
Library of Congress Control Number: 2016946468

SCITUS Academics LLC,
616, Corporate Way, Suite 2, 4766,
Valley Cottage, NY 10989

www.scitusacademics.com

Notice

Printed in United States of America on Acid Free Paper

Preface

Power quality is in fact the interaction between electricity and electrical equipment. If electrical equipment works reliably and correctly without harm or stress, we will say that electricity is of good quality. On the other hand, if the power equipment fails, is unreliable or is possibly damaged during regular use, we would think the quality of the power is poor. Any deviation from the normal voltage source (either AC or DC) can be classified as a power quality problem as a common statement. Power quality problems can be high-speed events such as voltage pulses/ transients, high-frequency noise, wave shape defects, voltage swells and total power losses. Every type of electrical equipment is affected by energy quality problems differently. By analyzing electricity and evaluating the equipment or possibly the load, we can decide whether there is a problem with the quality of power. Some common quality problems include rolling brow snouts, spikes, tension sags, harmonic distortion and electrical noise.

Power Quality Monitoring, Analysis and Enhancement contain various aspects of the monitoring, analysis and improvement of power quality in transmission and distribution systems. Electricity monitoring helps to identify important problems in power quality, such as voltage swells and swells, interruptions, harmonics and high-frequency noise, which are consistently seen in industrial and commercial grids. Fixing these problems requires accurate measurements and power quality analysis with monitoring instruments that can efficiently identify problems and identify solutions. This book will benefit experts, engineers and researchers.

Table of Contents

Chapter 1 Overview of power quality analysis and control technology for the smart grid 1

Chapter 2 Design of a Fixed-Order Robust Controller to Damp Inter-Area Oscillations in Power Systems 27

Chapter 3 Monitoring Power Quality in Small Scale Renewable Energy Sources Supplying Distribution Systems 51

Chapter 4 Hilbert-Huang transform with adaptive waveform matching extension and its application in power quality disturbance detection for microgrid 69

Chapter 5 HOS network-based classification of power quality events via regression algorithms 99

Chapter 6 Probabilistic Assessment of Power System Performance Quality 123

Chapter 7 Voltage sag compensation strategy for unified power quality conditioner with simultaneous reactive power injection 147

Chapter 8 Hybrid wind power balance control strategy using thermal power, hydro power and flow batteries 175

Chapter 9 **Study of Performance of Fresnel Lens Solar Concentrator** **219**

Chapter 10 **Link Quality-Based Transmission Power Adaptation for Reduction of Energy Consumption and Interference** **241**

Index **285**

Chapter 1

Overview of power quality analysis and control technology for the smart grid

An LUO[1] , Qianming XU[1] , Fujun MA[1] , Yandong CHEN[1]

National Electric Power Conversion and Control Engineering Technology Research Center (Hunan University), Changsha, China

ABSTRACT

With the wide application of non-linear loads and the large-scale access of distributed energy generations based on power electronics equipments, power quality problems in the distribution network are increasingly serious with new characteristics. Further in-depth research is of great significance in theory and practice. This paper provides an overview of power quality analysis, compensators, and control technologies under the new situation of smart grid. It focuses on the topologies and control methods for power quality conditioners, especially new characteristics of power quality and applicable control technologies in microgrids and distributed power plants. Finally, trends and prospects of power quality control technology are introduced, which is important to achieve security and efficient operation of the smart grid.

KEYWORDS

Power quality, Smart grid, Active control, Passive control

1. INTRODUCTION

In the modern smart grid, the diversity of loads and the demands for highly efficient consumption, as well as the use of renewable energy (solar, wind, biomass energy, etc.) generation and grid connection technology through the power electronics interfaces, have brought great challenges to governing power quality [1–4]. Compared with the traditional power system, the microgrid or distributed power plant, which integrates a variety of energy inputs, multiple load characteristics, and varied energy conversion technologies, is a nonlinear and complex system with inter-coupling of chemical energy, thermodynamics and electrodynamics. Meanwhile, due to the limitation of natural resources, the distributed generations (DGs) appear the features of intermittency, complexity, diversity and instability. Accordingly, some new problems with novel features occur in maintaining power quality. Therefore, power quality control theory and technology will play an important role in ensuring the stable and secure operation of the power grid when microgrids or distributed power plants are connected to it. In the Chinese national program for long- and medium-term scientific and technological development (2006-2020), it is specifically mentioned that the analysis, detection, and control technology for power quality should be a priority objective in the energy field, and it is highly recommended to promote the application of high-power power electronics technology and to control the power quality actively.

Figure 1 illustrates the typical structure of the smart grid including power quality controls. It consists of power transmission part, power distribution part, and power consumption part. The transmission grid is responsible for the regional power electricity supply and large renewable energy generation absorption. Its primary objective is to maintain the voltage stability of transmission lines by regulating appropriate reactive power provided by high voltage power quality compensators. The high- and medium-voltage distribution network works as the bridge that connects the transmission grid and power consumers. With the connection of large amounts of distributed energy, its power quality features appear high penetration and stochastic volatility. In consequence, multiple compensation devices are recommended to operate cooperatively for ensuring a friendly power quality. Power quality on the consumer side directly affects the power efficiency and production safety, which makes the power quality problems more complex. As a result, active control and local compensation technology are two important means to suppress the circulating current or harmonic resonance, and to analyze the emission features between DGs and electrical loads in the distribution network.

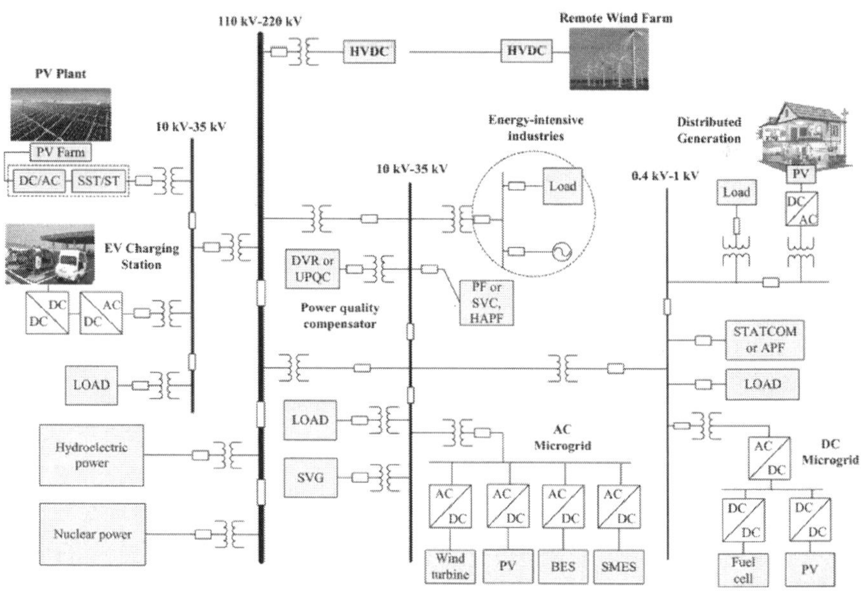

Figure 1. Structure diagram of a typical distribution grid system containing microgrid

This paper provides an overview of the power quality control technology in the smart grid. Power quality compensators and their classification based on power electronics technology are comprehensively introduced. A brief description is introduced for novel power quality compensators featuring modular multilevel technology, such as solid-state transformers and high voltage DC converters. Then, the control methods of power quality compensators are discussed for high efficiency operation. Special attention is paid to the new characteristics of power quality and control technology in the distribution system with DGs. Finally, the paper analyzes the trends and prospects of power quality control technology.

2. POWER QUALITY COMPENSATORS CLASSIFICATION

It is important to improve the power supply quality on power consumer side by taking advantage of the power quality control technology and equipment. The core principle of power quality control is to control and convert the electric energy, so as to meet the requirement of quality conformance and optimal efficiency. The key elements to accomplish above-mentioned control and conversion are diverse power electronics devices and associated control circuits. The classification of compensators in terms of different kinds of power quality problems is depicted in Fig. 2. Power quality control technology can be divided into active control technology and passive control technology.

2.1 Passive control technology

The passive control technology is characterized by adding extra devices to eliminate or relieve the impact of existing power quality problems. At present, the harmonic suppression techniques mainly contain the passive power filter (PPF) [5], active power filter (APF) [6, 7], and hybrid active power filter (HAPF) [8]. As mentioned in [9], HAPF can be classified as single-resonance injection hybrid active power filter (S-IHAPF), double-resonance injection hybrid active power filter (D-IHAPF), and multi-branch injection hybrid active power filter (M-IHAPF). The HAPF combines the advantages of both PPF and APF, effectively reducing the rated capacity and voltage of the active part without making a compromise of satisfactory performance in the integral filter system. It has been an effective way to suppress harmonic currents and compensate reactive power in high voltage distribution network. The reactive power compensator is capable of suppressing the voltage fluctuation and flicker. Off-the-shelf var compensators in distribution network include fixed capacitor (FC) [1, 2], static var compensator (SVC) [10], and static synchronous compensator (STATCOM) [11], etc. Among these devices, the STATCOM is widely used for its multiple functions, such as grid voltage fluctuation suppression, and unbalanced load compensation. The tendency has been to develop modularized harmonic suppression and var compensation systems due to high stability, reliability and power density.

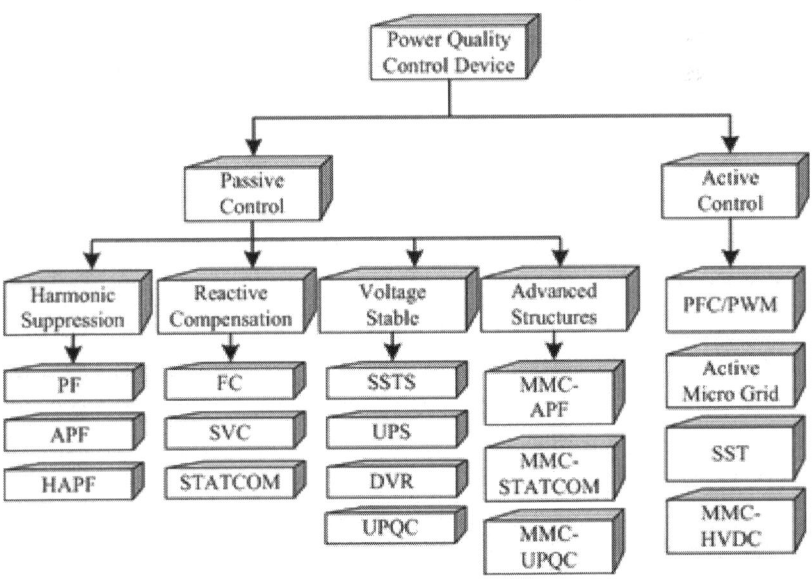

Figure 2. Classification diagram of power quality compensators

Among the transient power quality problems, voltage sag and short interruption are perceived as the most common and harmful forms. The solid-state transfer switch (SSTS) [12] can effectively reduce the depth and time of voltage sag. Uninterruptable power supply (UPS) [13] is the most effective tool to restrain the voltage fluctuation for low-power devices in the distribution network. The dynamic voltage regulator (DVR) [14] can directly and quickly compensate the instantaneous voltage sag and swell. The unified power quality controller (UPQC) [15], comprised of the series APF and shunt APF, can make comprehensive compensation of voltage and current for the grid.

In less than two decades, cascade power converters based on modular multilevel converter (MMC) have been widely studied in scientific researches and engineering applications. Due to the identical module structure, MMC greatly reduces the manufacturing difficulty and cost of high- and medium-voltage converters. Compared with the traditional power quality compensators, MMC based compensators have special advantages in standardization, expansibility, redundancy, fault ride-through, voltage level and filtering requirement [16, 17]. On the one hand, the modular power quality compensator can effectively lessen the rated voltage of the power switch and energy storage element in the sub module, which allows the use of low-loss and low-cost switch devices. On the other hand, the cascaded structure directly expands the application of diversified power quality compensator in medium- and high-voltage transmission systems. Currently, the APF, STATCOM, UPQC based on MMC have been discussed at a preliminary level [18–20]. Due to their novel topology, many new problems different from traditional power converters need to be addressed. More precisely, the multilevel structure inevitably requires a complex control system since large amounts of data must be processed within a short time. This limits the engineering application and development of MMC in the field of power quality control. More intensive theoretical and practical studies should be investigated to determine how to coordinate the interior power control, modulation method, circulating current suppression, and voltage balancing with exterior voltage and current compensation requirements. Fortunately, with the development of digital processing technology, the realization of complex topology and control system will gradually become simplified.

2.2 Active control technology

Active control technology is used to improve the impedance characteristics of the electrical equipment so as to prevent most power quality problems.

With the transmission and distribution network being instrumented, interconnected and intelligent, power quality problems caused by the future electrical equipment, especially the power electronics converting system, will be significantly mitigated. At present, the power factor correction (PFC) and pulse width modulation (PWM) technology have improved the power quality of rectifier devices [21]. The distributed generation and microgrid inverter using the active control [22] not only improve the quality of output voltage and current in distributed system, but also provide some extra compensation capacity for the adjacent grid. Further on, the prospective solid-state transformer (SST) [23] will block the transmission and emission of power quality problems between the power consumer side and power distribution side. The power quality level of the entire power grid will be enhanced due to the MMC-based high voltage direct current (HVDC) transmission and multi-terminal HVDC technology [24].

3. CONTROL METHOD OF POWER QUALITY COMPENSATOR

Nowadays, power quality control equipment generally employs the voltage source converter or current source converter. The output of converters can be regulated to control the power quality according to the voltage or current reference. The control method of the converter has a significant impact on the effect on power quality control. There are a lot of literatures describing the control methods, which are hysteresis control, dead-beat control, model predictive control, proportional integral control, proportional resonant control, repetitive control, and nonlinear robust control.

Hysteresis control (HC) is simple to realize and has a fast response speed, but its control accuracy depends on the hysteresis bandwidth and the switch frequency is unfixed [25,26]. Dead-beat control (DBC) is based on state equations of the converter and has a good dynamic performance with convenient parametric design, but its control performance is easily affected by circuit parameters and control delay. Compensating the control delay and adding state observation can strengthen the anti-jamming ability of the system [27–29]. Based on the finite control set of the converter, model predictive control (MPC) selects the output switch state to optimize the cost function. It features good dynamic performance, but it is also a challenge to the control system because of a large computation requirement [30–32]. However, developments in digital signal processing will contribute to the scale of application of MPC. In a synchronous reference coordinate system, proportional integral (PI) control [33, 34] has been widely used due to its tracking ability without error. Nevertheless, when applied to APF, it cannot realize zero error tracking over a broad frequency range. In order to improve the performance of alternating current control, proportional resonant (PR) control has been developed to achieve infinite gain and zero error tracking at the tuned frequency [35, 36]. Repetitive control, derived from the internal model principle, makes it possible to accurately track the periodic signal and enhance the steady-state performance [37, 38], but its dynamic response still needs to be improved. Making use of the nonlinear characteristics of the converter, sliding mode control based on feedback linearization is proposed in [39–41], where it can improve the dynamic performance as well as the robustness of the system. In [42, 43], the artificial neural network controller and the linear quadratic Gaussian controller are designed to optimize the performance of APF.

It is noteworthy that a lot of literatures promote methods to improve the compensation performance. [44] presents a multi-PR current controller based on the circuit model, which simplifies the harmonic extraction unit. [45] describes a control method with decoupled fundamental component and harmonic component, resulting in dispensable unit for the harmonic voltage and current detection. Dividing frequency control proposed in [8] and [46] reduces the compensation error of the harmonic current's amplitude and phase, which is suitable for multi-structure active power filters.

4. ANALYSIS AND CONTROL OF POWER QUALITY FOR LARGE SCALE DISTRIBUTED POWER PLANTS

With the increasing penetration rate of large-scale distributed power plants by year, the cross-coupling of harmonic resonance between the distributed system and multi-inverter system has become increasingly complicated. The distributed power plant generates a lot of high-frequency and broadband harmonic components [47]. Moreover, there are series and parallel resonances depending on the transmission line configuration and harmonic frequencies. In cases where the background harmonic voltage matches the transmission line parameters, there will be series resonance, resulting in serious amplification of harmonic voltage. Harmonic currents generated by distributed power plants will also result in parallel resonance, causing serious current distortion, further increasing the total harmonic content in the system. Series–parallel resonances are mainly responsible for voltage and current distortions, which disturb the efficient and stable operation of distributed power plants [48, 49]. Therefore, analyzing series and parallel resonance mechanisms of distributed power plant has become a hot topic in power system research.

The influence of transmission line distributed capacitance on harmonic series and parallel resonance is analyzed in [50]. Then, a preliminary resonance suppression method is proposed in [51]. Quantitative analyses of the resonance caused by transformer magnetizing branch, parallel capacitors and other devices are presented in [52]. In [53], the step-up transformer, the reactive power compensation device, distributed capacitance of the transmission line and loads are considered to establish the multiport passive network model. This allows the transmission mechanism of harmonics to be analyzed using the definition of resonance amplification factors. As shown in Fig 3a, double resonance curves are discovered, which means that transmission lines of different distances may amplify a harmonic current twice, or a transmission line of a fixed distance may amplify the harmonic current of different frequencies. For transmission line distance within 100 km in resonance curve 1, the harmonics resonance frequency is between the 11th and 27th, and the amplification factor for the 17th and 19th harmonics may reach 20. In resonance curve 2, the dominant resonance frequency within 200–300 km transmission line is the 9th harmonic, and its amplification factor can reach 10 at the resonance point. As depicted in Fig 3b, apparently, in the case of long-distance transmission (100–300 km), the most dominant resonance frequency is between the 5th and 15th harmonics; the 7^{th}harmonic voltage is magnified about 20 times. Whereas in the case of short-distance transmission (0–100 km), the dominant series resonance frequency is between the $17^{th}$and 50th harmonics. At the resonance points, the 17th and 19th harmonic voltages may magnify more than 20 times, and more than 100 times for the 23rd, 27th, 37th and 49^{th}harmonic voltages.

Harmonics generated by large-scale distributed power plants have the characteristics of high frequency and wide frequency range. Therefore, the adverse impact of distributed parameters of high-voltage cables becomes more significant [54].

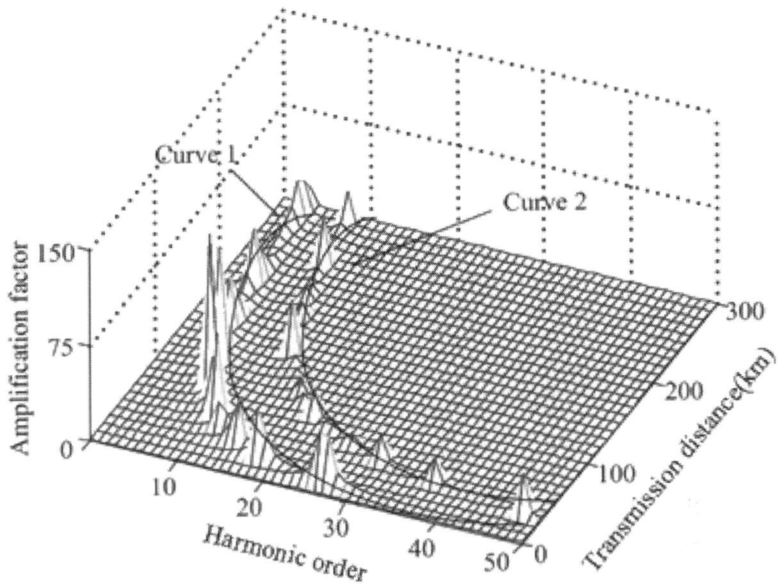

(a) Harmonic current amplification factor

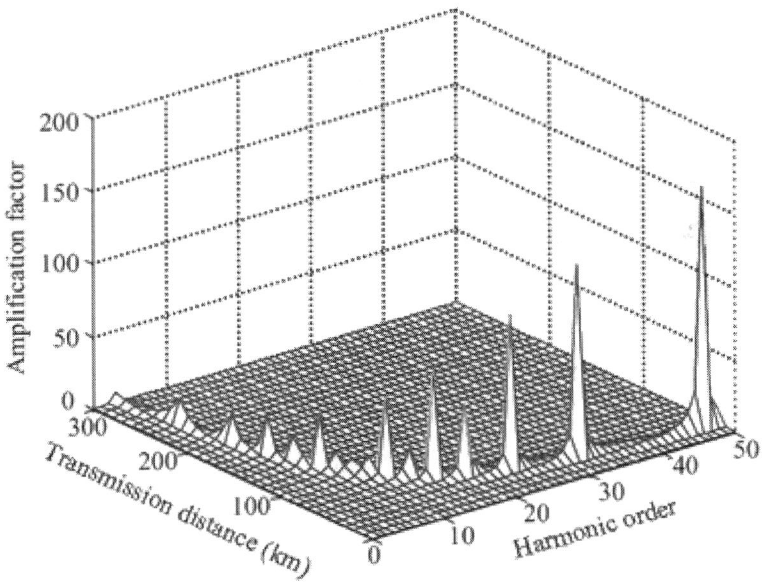

(b) Harmonic voltage amplification factor

Figure 3. The relationship between the resonance amplification factor and harmonic order, transmission distance

The methods to eliminate the series-parallel harmonic resonance can be summarized in two categories:

1) By changing the parameters of the transmission network, the harmonic resonance can be restrained [55]. Existing measures set the shunt reactors at both ends of transmission lines to weaken the effect of distributed capacitances. But this approach shows some drawbacks when it is applied to distributed power plants. On the one hand, the shunt reactors reduce the power factor at the export side of distributed power plants. On the other hand, the variation of the output active and reactive power of distributed power plants may cause large voltage fluctuations at the point of common coupling (PCC).

2) Adding filter device can markedly reduce the harmonic content flowing into the grid [56]. Most conventional filter devices are available for lower order harmonic currents compensation, but incapable of eliminating harmonic frequencies more than 25th. Hence, new filter topologies that can effectively accomplish broadband harmonic detection and elimination will undoubtedly be paid critical attention in the near future.

It is worthy of note that the hybrid active power filter with the grid current detection function [57] can be equivalent to a harmonic impedance in series with the grid impedance. It can ameliorate the grid equivalent impedance to suppress the resonance. Concurrently, harmonic currents will be filtered at low cost. Hence, it is useful for the harmonic management of distributed power plants.

5. ANALYSIS AND CONTROL OF POWER QUALITY FOR MICROGRIDS AND POWER DISTRIBUTION SYSTEMS CONTAINING MICROGRIDS

The DG units, energy storage system (ESS) and loads are connected to the microgrid through power electronics converters. This structure is the main form of the DGs for renewable energy at present. The power quality problems of the microgrid exist in a wide-band frequency domain along with an increasing penetration of the microgrid. Determining how to effectively tackle the problem of the harmonic currents in the wide-band frequency domain caused by the distributed and clustered inverters has been gained much attention in the flied of the power quality of the microgrid operating in both islanded and grid-connected mode [58]. The most significant problems in the islanded mode are how to effectively realize power sharing among parallel inverters, also reduce the fundamental and harmonic circulating currents among them. The prominent problems in grid-connected mode are how to solve the coupling resonant among parallel inverters as well as between the inverters and the distribution network. This section will focus on the new characteristics of the power quality and its control technologies in the microgrid.

5.1 Suppression methods of internal circulating current in microgrids

The equivalent output impedances of the parallel inverters are different from each other at PCC [59] in an AC or hybrid microgrid since the control strategies, filter parameters and length of transmission lines [60], are not the same. Consequently, it will cause the fundamental and harmonic circulating currents, and also the fluctuations of the amplitude and frequency of the PCC voltage of the islanded microgrid [61, 62]. There have three types of circulating currents in the islanded mode, i.e., active circulating current, reactive circulating current and harmonic circulating current. The active circulating current is caused by the frequency and phase deviations of inverter output voltages. The reactive circulating current results from the magnitude deviations of inverter output voltages. The harmonic circulating current is the result of harmonic components of inverter output voltages. The slight deviations of the frequency, phase and magnitude of the inverter output voltages will lead to large fundamental or harmonic circulating currents [63]. Moreover, the circulating currents will reduce the output capacities of the DG units, and increase the line losses, also endanger the safe operation of the microgrid [64, 65].

The virtual impedance technology can effectively reduce the circulating currents by changing the equivalent output impedance of the parallel inverters. The resistive (R-type) inverter, inductive (L-type) inverter, resistive-inductive (RL-type) inverter, and resistive-capacitive (RC-type) inverter can be easily designed by using the virtual impedance technology to meet the different operating requirements of the microgrid [66–69]. In general, the output impedance of the inverters in the middle-voltage-level microgrid is designed to be inductive; on the contrary, the output impedance is often designed to be resistive-inductive in the low-voltage-level microgrid.

5.2 Harmonic analysis and suppression in the microgrids operating in grid-connected mode

The nonlinear action and dead time effect of distributed inverter bring plenty of high-order harmonic currents in the microgrids. Thus, the output filter devices for inverters are indispensable. There have two main types of output filter for inverters in grid-connected mode, i.e., L-type and LCL-type filter. Besides, the LC-type filter is mainly used in the islanded mode [70]. The equivalent output impedance of each inverter will easily be changed due to the deviation of filter parameters. Impedance coupling exists among the equivalent impedances of the parallel inverters, as well as between the equivalent output impedance of the inverter and the equivalent impedance of the distribution network. Therefore, it can provide a low-impedance channel for specific harmonic, resulting in the amplification of the harmonic current with specific orders. The active damping loop can provide solutions to this problem in terms of the inverter control. Cascade control method [71, 72], splitting capacitance method [73], zero-pole compensation method

[74], and filter-capacitor-voltage feedback method [75] are the effective active damping schemes.

The resonant problem of multi-parallel inverters is complex [76, 77], and there exists both a fixed and an unfixed resonant band. The fixed resonant band is caused by inverter filter system, and its resonant frequencies will not migrate with the different number of parallel inverters. However, the unfixed resonant band results from the coupling effect of the equivalent output impedance of the parallel inverter system and its resonant frequencies will migrate to the low-frequency range along with the increase in the number of the parallel inverters. In addition, the series resonance exists between the parallel inverter and the distribution network [78], and its resonance peak frequency will also migrate to the low-frequency range along with the increase in the number of parallel inverters. To restrain the series resonance, the Nyquist stability criterion can be extended to provide a feasible solution for reshaping the equivalent output impedance of the inverter. Besides, the active damping schemes should be taken to attenuate the coupling resonance among parallel inverters.

6. TRENDS AND PROSPECTS OF POWER QUALITY CONTROL

With concern about the consumption of fossil fuels and the resulting greenhouse effect, a large number of sustainable energy sources are connected to the utility grid in the form of DGs. This will affect the power quality of power grid and bring new demands of the compensation devices. At the same time, the potential application of wide band semiconductors (SiC), and the development of new converter topologies will promote the development of advanced power quality compensators. In addition, solid state circuit breakers, SSTS and MMC-HVDC, which can cooperatively control active power and reactive power to achieve power quality active control, have been extensively studied.

The continuous application of such technology will have a positive impact on power quality control.

Considering the issues and technologies presented in these papers, the following areas require further research:

1) The harmonics generated by power electronics converters have characteristics of high-frequency, wide frequency range, and multi-type coupling. How to detect and suppress the high-order harmonics and inter-harmonics still needs to be further investigated.

2) The emission level and superposition characteristic of the power quality problem as well as the interaction effect between the micro-grid and the utility grid are worth researching.

3) Power quality characteristics of large wind power farms, photovoltaic power plants, their interaction effect with public power grid and suitable power quality control methods merit intensive research.

4) The characteristics of the power quality in large scale distributed power plant as well as the power quality control method are prime candidates for in-depth study.

5) New designs for power quality compensators based on wide band semiconductors or modular multilevel technology need to be developed.

6) The functionality and widespread application of solid state circuit breakers, solid state transformers and MMC-HVDC, all of which can perform active power quality control, require in-depth research.

7. CONCLUSION

With the growth of extensive nonlinear loads and renewable power integration, power quality problems in distribution grids have become increasingly serious. Problems such as voltage fluctuation, voltage sag, harmonic and reactive power, not only directly harm the normal operation of sensitive loads, but also decrease the power transmission efficiency of the distribution system. This paper summarizes available and emerging control technologies to solve power quality problems in the strong smart grid. Particularly, it focuses on new power quality features, control technologies and trends and prospects, to form a valuable reference to assist the secure and economical operation of distribution grids.

ACKNOWLEDGMENT

This work was supported by the General Program of National Natural Science Foundation of China (No. 51477045).

REFERENCES

1. Miller TJE (1982) Reactive power control in electric system. Wiley, New York

2. Wang ZA, Yang J, Liu JJ (2006) Harmonic suppression and reactive power compensation. Machinery Industry Press, Beijing **(in Chinese)**

3. Green MA (2002) Third generation photovoltaics: comparative evaluation of advanced solar conversion options. In: Proceedings of the 29th IEEE photovoltaic specialists conference, New Orleans, 19–24 May 2002, pp 1330–1334

4. Zhao ZM, Liu JZ, Sun XY (2005) Solar photovoltaic power generation and its application. Science Press, Beijing **(in Chinese)**

5. El-Saadany EF, Salama MMA, Chikhani AY (2000) Passive filter design for harmonic reactive power compensation in single-phase circuits supplying nonlinear loads. IEE P-Gener Transm Distrib 147(6):373–380

6. Akagi H (1994) Trend in active power line conditioners. IEEE Trans Power Electron 9(3):263–268

7. Khadkikar V, Chandra A (2008) A new control philosophy for a unified power quality conditioner (UPQC) to coordinate load-reactive power demand between shunt and series inverters. IEEE Trans Power Deliv 23(4):2522–2534

8. Luo A, Zhao W, Deng X et al (2009) Dividing frequency control of hybrid active power filter with multi-injection branches using improved i_p-i_q algorithm. IEEE Trans Power Electron 24(10):2396–2405

9. Zhao W (2010) Harmonic suppression and reactive power compensation theory and application research in high voltage distribution networks. Ph D Thesis, Hunan University, Changsha (**in Chinese**)

10. Gyuyi L, Talor ER (1980) Characteristic of static, thyristor-controlled shunt compensators for power transmission system applications. IEEE Trans Power Appar Syst 99(5):1795–1804

11. Ghosh A, Ledwich G (2003) Load compensating DSTATCOM in weak AC systems. IEEE Trans Power Deliv 18(4):1302–1309

12. Chan K, Kara A, Kieboom G (1998) Power quality improvement with solid state transfer switches. In: Proceedings of the 8th international conference on harmonics and quality of power, vol 1, Athens, 14–18 Oct 1998, pp 210–215

13. Bashi SM, Jasni J, Weng OY (2003) Voltage regulation of uninterrupted power supplies. In: Proceedings of the IEEE student conference on research and development (SCOReD'03), Putrajaya, 25–26 Aug 2003, pp 385–389

14. Liu JW, Choi SS, Chen S (2003) Design of step dynamic voltage regulator for power quality enhancement. IEEE Trans Power Deliv 18(4):1403–1409

15. Fujita H, Akagi H (1998) The unified power quality conditioner: the integration of series and shunt-active filters. IEEE Trans Power Electron 13(2):315–322

16. Lesnicar A, Marquardt R (2003) An innovative modular multilevel converter topology suitable for a wide power range. In: Proceedings of the 2003 IEEE Bologna power technology conference, vol 3, Bologna, 23–26 June 2003, 6 pp

17. Kouro S, Malinowski M, Gopakumar K et al (2010) Recent advances and industrial applications of multilevel converters. IEEE Trans Ind Electron 57(8):2553–2580

18. Ghetti FT, Ferreira AA, Brage HAC, et al. (2012) A study of shunt active power filter based on modular multilevel converter (MMC). In: Proceedings of the 10th IEEE/IAS international conference on industry applications (INDUSCON'12), Fortaleza, 5–7 Nov 2012, 6 pp

19. Nieves M, Maza JM, Mauricio JM, et al. (2014) Enhanced control strategy for MMC-based STATCOM for unbalanced load compensation. In: Proceedings of the 16th European conference on power electronics and applications (EPE'14-ECCE Europe), Lappeenranta, 26–28 Aug 2014, 10 pp

20. Long YB, Xiao XN, Xu YH et al (2013) MMC-UPQC: application of modular multilevel converter on unified power quality conditioner. In: Proceedings of the 2013 IEEE Power and Energy Society general meeting (PES'13), Vancouver, 21–23 July 2013, 5 pp

21. Yao K, Ruan XB, Zou C et al (2012) Three-phase single-switch boost PFC converters with high input power factor. Proc CSEE 32(6):97–105 **(in Chinese)**

22. Zeng Z, Zhao RX, Yang H et al (2012) A multi-functional grid-connected inverter and its application to customized power quality of microgrid. Power Syst Technol 36(5):58–67 **(in Chinese)**

23. Leung CK, Dutta S, Baek S et al (2010) Design considerations of high voltage and high frequency three phase transformer for solid state transformer application. In: Proceedings of the 2010 IEEE energy conversion congress and exposition (ECCE'10), Atlanta, 12–16 Sept 2010, pp 1551–1558

24. Railing RD, Moreau G, Ronström L et al (2004) Cross sound cable project second generation VSC technology for HVDC. In: Proceedings of the 2004 CIGRE conference, Paris, 21–23 July 2004, pp B4–B102

25. Lam CS, Wong MC, Han YD (2012) Hysteresis current control of hybrid active power filters. IET Power Electron 5(7):1175–1187

26. Suresh Y, Panda AK, Suresh M (2012) Real-time implementation of adaptive fuzzy hysteresis-band current control technique for shunt active power filters. IET Power Electron 5(7):1188–1195

27. Kamran F, Habetler TG (1998) Combined deadbeat control of a series-parallel converter combination used as a universal power filter. IEEE Trans Power Electron 13(1):160–168

28. Han Y, Xu L, Khan MM et al (2011) Robust deadbeat control scheme for a hybrid APF with resetting filter and ADELINE-based harmonic estimation algorithm. IEEE Trans Ind Electron 58(9):3893–3904CrossRef

29. Biagini V, Zanchetta P, Odavic M et al (2013) Control and modulation of a multilevel active filtering solution for variable-

speed constant-frequency more-electric aircraft grids. IEEE Trans Ind Inform 9(2):600–608

30. Drobni K, Nemec M, Nedeljkovic D et al (2009) Predictive direct control applied to AC drives and active power filter. IEEE Trans Ind Electron 56(6):1884–1893

31. Acuña P, Morán L, Rivera M et al (2015) A single-objective predictive control method for a multivariable single-phase three-level NPC converter-based active power filter. IEEE Trans Ind Electron 62(7):4598–4607

32. Panigrahi R, Subudhi B, Panda PC (2015) Model predictive-based shunt active power filter with a new reference current estimation strategy. IET Power Electron 8(2):221–233

33. Xu Y, Li FX (2014) Adaptive PI control of STATCOM for voltage regulation. IEEE Trans Power Deliv 29(3):1002–1011

34. Rahmani S, Mendalek N, Al-Haddad K (2010) Experimental design of a nonlinear control technique for three-phase shunt active power filter. IEEE Trans Ind Electron 57(10):3364–3375

35. Micallef A, Apap M, Spiteri-Staines C et al (2014) Reactive power sharing and voltage harmonic distortion compensation of droop controlled single phase islanded microgrids. IEEE Trans Smart Grid 5(3):1149–1158

36. Herman L, Papic I, Blazic B (2014) A proportional-resonant current controller for selective harmonic compensation in a hybrid active power filter. IEEE Trans Power Deliv 29(5):2055–2065

37. Zou ZX, Zou KL, Wang Z et al (2015) Frequency-adaptive fractional-order repetitive control of shunt active power filters. IEEE Trans Ind Electron 62(3):1659–1668

38. Grino R, Cardoner R, Costa-Castelló R et al (2007) Digital repetitive control of a three-phase four-wire shunt active filter. IEEE Trans Ind Electron 54(3):1495–1503

39. Wu JY, Chen QC, Du MJ et al (2013) Sliding-mode variable structure controller for cascade static var compensator. IET Trans Power Electron 6(2):343–352

40. Matas J, de Vicuña LG, Miret J et al (2008) Feedback linearization of a single-phase active power filter via sliding mode control. IEEE Trans Power Electron 23(1):116–125

41. Al Chaer T, Gaubert JP, Rambault L et al (2009) Linear feedback control of a parallel active harmonic conditioner in power systems. IEEE Trans Power Electron 24(3):641–653

42. Radzi MAM, Rahim NA (2009) Neural network and bandless hysteresis approach to control switched capacitor active power filter for reduction of harmonics. IEEE Trans Ind Electron 56(5):1477–1484

43. Panigrahi R, Subudhi B, Panda PC (2016) A robust LQG servo control strategy of shunt- active power filter for power quality enhancement. IEEE Trans Power Electron 31(4):2860–2869

44. He JW, Li YW, Blaabjerg F et al (2014) Active harmonic filtering using current-controlled, grid-connected DG units with closed-loop power control. IEEE Trans Power Electron 29(2):642–653

45. Zobaa AF (2014) Optimal multiobjective design of hybrid active power filters considering a distorted environment. IEEE Trans Ind Electron 61(1):107–114

46. Luo A, Shuai ZK, Zhu WJ et al (2012) Combined system for harmonic suppression and reactive power compensation. IEEE Trans Ind Electron 56(2):418–428

47. Agorreta JL, Borrega M, López J et al (2011) Modeling and control of *N*-paralleled grid-connected inverters with LCL filter coupled due to grid impedance in PV plants. IEEE Trans Power Electron 26(3):770–785

48. Xu W, Huang ZY, Cui Y et al (2005) Harmonic resonance mode analysis. IEEE Trans Power Deliv 20(2):1182–1190

49. Xu DZ, Wang F, Mao HL et al (2013) Modeling and analysis of harmonic interaction between multiple grid-connected inverters and the utility grid. Proc CSEE 33(12):64–71 **(in Chinese)**

50. Shuai ZK, Liu DG, Shen J et al (2014) Series and parallel resonance problem of wideband frequency harmonic and its elimination strategy. IEEE Trans Power Electron 29(4):1941–1952

51. Shuai ZK, Xiao F, Tu CM et al (2013) Resonance degradation mechanism of wide-band frequency harmonics and elimination strategy. Trans China Electrotech Soc 28(12):16–23 **(in Chinese)**

52. Zhong HQ, Xu ZX, Zou YP et al (2005) Effects of parasitical capacitors on charging characteristic of series resonant CCPS. Proc CSEE 25(10):40–44 **(in Chinese)**

53. Luo A, Xie N, Shuai ZK (2015) Large-scale photovoltaic plant harmonic transmission model and analysis on resonance characteristics. IET Power Electron 8(4):565–573

54. Infield DG, Onions P, Simmons AD et al (2004) Power quality from multiple grid-connected single-phase inverters. IEEE Trans Power Deliv 19(4):1983–1989

55. Indulkar CS, Viswanathan B, Venkata SS (1989) Maximum power transfer limited by voltage stability in series and shunt compensated schemes for AC transmission systems. IEEE Trans Power Deliv 4(2):1246–1252

56. Ma FJ, Luo A, Shuai ZK et al (2014) Voltage ripple analysis of simplified active power compensator for negative sequence and reactive power compensation. IET Power Electron 7(10):2582–2594

57. Luo A, Peng SJ, Wu CP et al (2012) Power electronic hybrid system for load balancing compensation and frequency-selective harmonic suppression. IEEE Trans Ind Electron 59(2):723–732

58. Li YW, Nejabatkhah F (2014) Overview of control, integration and energy management of microgrids. J Mod Power Syst Clean Energy 2(3):212–222.

59. Guerrero JM, De Vicuna LG, Matas J et al (2005) Output impedance design of parallel-connected UPS inverters with wireless load-sharing control. IEEE Trans Ind Electron 52(4):1126–1135

60. Tuladhar A, Jin H, Unger T et al (2000) Control of parallel inverters in distributed AC power systems with consideration of line impedance effect. IEEE Trans Ind Appl 36(1):131–138

61. Guerrero JM, De Vicuna LG, Matas J et al (2004) A wireless controller to enhance dynamic performance of parallel inverters in distributed generation systems. IEEE Trans Power Electron 19(5):1205–1213

62. Ding GQ, Gao F, Zhang S, Loh PC, Blaabjerg F (2014) Control of hybrid AC/DC microgrid under islanding operational conditions. J Mod Power Syst Clean Energy 2(3):223–232.

63. Vasquez JC, Mastromauro RA, Guerrero JM et al (2009) Voltage support provided by a droop-controlled multifunctional inverter. IEEE Trans Ind Electron 56(11):4510–4519

64. De Brabandere KD, Bolsens B, Van den Keybus J et al (2007) A voltage and frequency droop method for parallel inverters. IEEE Trans Power Electron 22(4):1107–1115

65. Zhong QC (2013) Robust droop controller for accurate proportional load sharing among inverters operated in parallel. IEEE Trans Ind Electron 60(4):1281–1290

66. Yao W, Chen M, Matas J et al (2011) Design and analysis of the droop control method for parallel inverters considering the impact of the complex impedance on the power sharing. IEEE Trans Ind Electron 58(2):576–588

67. Guerrero JM, Vasquez JC, Matas J et al (2011) Hierarchical control of droop-controlled AC and DC microgrids: a general approach towards standardization. IEEE Trans Ind Electron 58(1):158–172

68. Zhong QC, Zeng Y (2014) Control of inverters via a virtual capacitor to achieve capacitive output impedance. IEEE Trans Power Electron 29(10):5568–5578

69. Chen YD, Guerrero JM, Shuai ZK et al (2015) Fast reactive power sharing, circulating current and resonance suppression for parallel inverters using resistive-capacitive output impedance. IEEE Trans Power Electron.

70. Wai RJ, Lin CY, Huang YC et al (2013) Design of high-performance stand-alone and grid-connected inverter for distributed generation applications. IEEE Trans Ind Electron 60(4):1542–1555

71. Twining E, Holmes DG (2003) Grid current regulation of a three-phase voltage source inverter with an LCL input filter. IEEE Trans Power Electron 18(3):888–895

72. Dannehl J, Fuchs FW, Hansen S et al (2010) Investigation of active damping approaches for PI-based current control of grid-connected pulse width modulation converters with LCL filters. IEEE Trans Ind Appl 46(4):1509–1517

73. Shen GQ, Xu DH, Cao LP et al (2008) An improved control strategy for grid-connected voltage source inverters with an LCL filter. IEEE Trans Power Electron 23(4):1899–1906

74. Malinowski M, Bernet S (2008) A simple voltage sensorless active damping scheme for three-phase PWM converters with an LCL filter. IEEE Trans Ind Electron 55(4):1876–1880

75. Dannehl J, Liserre M, Fuchs FW (2011) Filter-based active damping of voltage source converters with LCL filter. IEEE Trans Ind Electron 58(8):3623–3633

76. Wang F, Duarte JL, Hendrix MAM et al (2011) Modeling and analysis of grid harmonic distortion impact of aggregated DG inverters. IEEE Trans Power Electron 26(3):786–797

77. Ghiani E, Pilo F (2015) Smart inverter operation in distribution networks with high penetration of photovoltaic systems. J Mod Power Syst Clean Energy 3(4):504–511.

78. Enslin JHR, Heskes PJM (2004) Harmonic interaction between a large number of distributed power inverters and the distribution network. IEEE Trans Power Electron 19(6):1586–1593

Chapter 2

Design of a Fixed-Order Robust Controller to Damp Inter-Area Oscillations in Power Systems

Abdlmnam Abdlrahem[1], Parimal Saraf[1], Karthikeyan Balasubramaniam[1], Ramtain Hadidi[1], Alireza Karimi[2], Elham Makram[1]

[1]Department of Electrical and Computer Engineering, Clemson University, Clemson, USA

[2]Laboratoired'Automatique, Ecole Polytechnique Fédérale de Lausanne (EPFL), Lausanne, Switzerland

ABSTRACT

This paper presents the design of a robust fixed-order H_∞ controller to damp out the inter-area oscillations and to enhance the stability of the power system. The proposed H_∞approach is based on shaping the open-loop transfer function in the Nyquist diagram through minimizing the quadratic error between the actual and the desired open loop transfer functions in the frequency domain under linear constraints that guarantee robustness and stability. The proposed approach is robust with respect to multi-model uncertainty closed-loop sensitivity functions in the Nyquist diagram through the constraints on their infinity norm. The H_∞ constraints are linearized with the help of a desired open-loop transfer function. The controller is designed using the convex optimization techniques in which the difference between the open-loop transfer function and the desired one is minimized. The two-area four-machine test system is selected to evaluate the performance of the designed controller under different load conditions as well as different levels of wind penetrations.

KEYWORDS

H_∞, Multi-Machine Power System, Nyquist Diagram, Robust Control, Wind Penetrations, SVC

1. INTRODUCTION

The rapid growth of the power system has been added more challenges to the capability of power transfer between interconnected areas. Damping of inter-area oscillations (0.2 Hz to 1 Hz) is considered as one of the main challenges for the interconnected systems [1] . These low frequency oscillations may grow and lead to the loss of system stability. Recently, Flexible AC Transmission System (FACTS) devices are installed in power systems to improve the voltage profile and tie-line power. The damping of the overall system cannot be improved using these devices alone. Supplementary control signals are required to be added to these devices to achieve more damping [2] - [4] .

In this work, the main objective of control design is to ensure sufficient damping under different power flow conditions. Different approaches have been investigated to damp power system oscillations. Recently, many studies have investigated using of H_2, H_∞ optimization [5] , and μ-synthesis [6] for designing a robust controller for power systems. The aim of these approaches is to damp the power system oscillations and enhance the stability of power system.

The order of the designed controller has to be as low as possible since the controller needs to be implemented in computers that have limited memory and computing power. In the above approaches, the power system has to be reduced to lower order and that cannot guarantee that the reduced order controller will achieve the requirements of stability and performance.

In this paper a fixed-order linearly parameterized controller is designed using the H_∞ approach. The main idea of the proposed approach is based on shaping of the open-loop transfer function under an infinite number of convex constraints on the Nyquist diagram, [7] . Considering the multi-model uncertainty is the main advantage of this approach. In addition, it doesn't suffer from the drawback that most methods have which is reducing the order of the plant. This method directly deals with the entire plant and still leads to a low order controller. The proposed approach can handle both stable and unstable plant models. In this work, however, only stable plant models are considered. Frequency Domain Robust Control (FDRC) Toolbox which is introduced in [7] is used in this paper to design H_∞ fixed-order robust controller.

The rest of this paper is organized as follows: designing linearly parameterized H_∞ robust controller in the Nyquist diagram is recalled in Section 2. Section 3 explains the test power system (two-area four-machine system) including Static VAR Compensator (SVC) and wind models. Section 4 describes the controller design, the results of the time domain simulations and the eigenvalue study. Finally, Section 5 provides concluding remarks.

2. H∞ CONTROLLER

The primary purpose of this paper is to design linearly parameterized robust controller to damp out the inter-area oscillations. Consider a linearly parameterized controller of the form given in (1) [8] [9] :

$$K(s)=\rho^{T}\varphi(s) \tag{1}$$

where,

$$\rho=[\rho_1\rho_2\cdots\rho_n]$$

$$\varphi(s)=[\varphi_0(s)\varphi_1(s)\cdots\varphi_{n-1}(s)]^{T}$$

and n is the number of controller parameters, ρ_i is the controller parameters and $\varphi_i(s)$ is a basis function. The Laguerre function is a commonly used basis function and is given in (2).

$$\varphi_0(s)=1,\ \varphi_i(s)=\frac{\sqrt{2\zeta}\,(s-\zeta)^{i-1}}{(s+\zeta)^{i}},\ i\geq 1, \zeta>0 \tag{2}$$

where $\zeta>0$ is the Laguerre parameter. It can be shown that for any finite order transfer function F(s), arbitrary Laguerre parameter $\zeta>0$ and an arbitrary constant $\varepsilon>0$, there exists a finite n such that

$$\left\|F(s)-\rho^{T}\varphi(s)\right\|_{p}<\varepsilon \quad \text{for } 0<p<\text{infinity} \tag{3}$$

The controller parameterization presented in (1) allows us to get a good approximation of any finite order stable transfer function with a desired level of accuracy by varying the parameter n. The result of the optimization problem given in (3) is dependent on the difference between the poles of F(s) and ζ. A better approximation of any finite order stable transfer function can be obtained for a given controller order if the choice of ζ is proper. More details for optimal selection of the basis function can be found in [8] .

The reason behind using the linearly parameterized controller is because all points on the Nyquist diagram of the open loop transfer function $L(j\omega,\rho)$ can be written as a linear function of the controller parameters ρ as given in (4). This property helps in obtaining a convex parameterization of the H_∞ fixed order controller.

$$L(j\omega,\rho)=K(j\omega,\rho)G(j\omega)=\rho^{T}\varphi(j\omega)G(j\omega)=\rho^{T}\mathcal{R}(\omega)+j\rho^{T}\mathcal{J}(\omega) \quad (4)$$

where $\mathcal{R}(\omega)$ and $\mathcal{J}(\omega)$ are respectively the real and imaginary parts of $\varphi(j\omega)G(j\omega)$.

In case of a single model, G is a scalar function whereas for a multi-model controller design $\mathcal{G}=\{G_i(j\omega),i=1,\cdots,m\}$ is defined where $G_i(j\omega)$ represents the i-th model in the multi-model uncertainty set. In this case, $L_i(j\omega)$is the open-loop transfer function for the i-th model.

2.1. The Robust Performance Constraints

The main idea of this method is based on the shaping of the open-loop transfer function under infinite number of convex constrains on the Nyquist diagram such that the following constrain is satisfied:

$$|W_1(j\omega)S(j\omega)|+|W_2(j\omega)T(j\omega)|<1 \quad \forall\omega \quad (5)$$

where $T=L(j\omega)/[1+L(j\omega)]$ is the complementary sensitivity function and $S=1/[1+L(j\omega)]$ is the sensitivity function.

The main concept is to represent the constraints in (5) in the Nyquist plot then the robustness can be achieved by a set of convex constraints on the Nyquist plot. Now, we can represent the controller design by a convex optimization problem and the aim is to minimize the norm of the difference between the actual and desired open- loop transfer function.

Multiplying the constraints in (5) by $|1+L(j\omega,\rho)|$, one obtains:

$$|W_1(j\omega)|+|W_2(j\omega)L(j\omega,\rho)|<|1+L(j\omega,\rho)| \quad \forall\omega \tag{6}$$

This approach is based on minimizing the difference between the desired open loop transfer function $L_d(j\omega_k)$ and the open loop transfer function $L_i(j\omega_k,\rho)$ shown in Figure 1 [8] . In the first step the non-convex constraints in (6) are linearized around the desired open-loop transfer function.

As it is well known, $(-1+j0)$is the critical point on the Nyquist plot for analyzing the stability of the closed-loop system. The distance between the critical point and the open-loop transfer function is $|1+L(j\omega,\rho)|$. In Figure 1 the robust performance condition in (6) is met if and only if the Nyquist plot of a circle with radius $|W_1(j\omega_k)|$ centered at $(-1+j0)$ does not intersect with a circle (uncertainty circle) centered at $L_i(j\omega_k,\rho)$ with radius $|W_2(j\omega_k)L(j\omega_k,\rho)|$ for all ω [9] . The line d as shown in Figure 1 is tangent to the circle cen- tered at the critical point and orthogonal to the line that connects the center of the critical point to the desired open-loop transfer function $L_d(j\omega_k)$. A sufficient condition for (6) is that the uncertainty circle lies at the side of d that excludes the critical point for all frequencies. The line d has infinite number of points and each point can be represented as a point in the complex plane, has real part x and imaginary part y. At each frequency ω (each point) the equation of the line d can be written as:

$$\text{line } d: y=\tan(\alpha)\left[x-\frac{M}{\sin(\alpha)}+1\right] \tag{7}$$

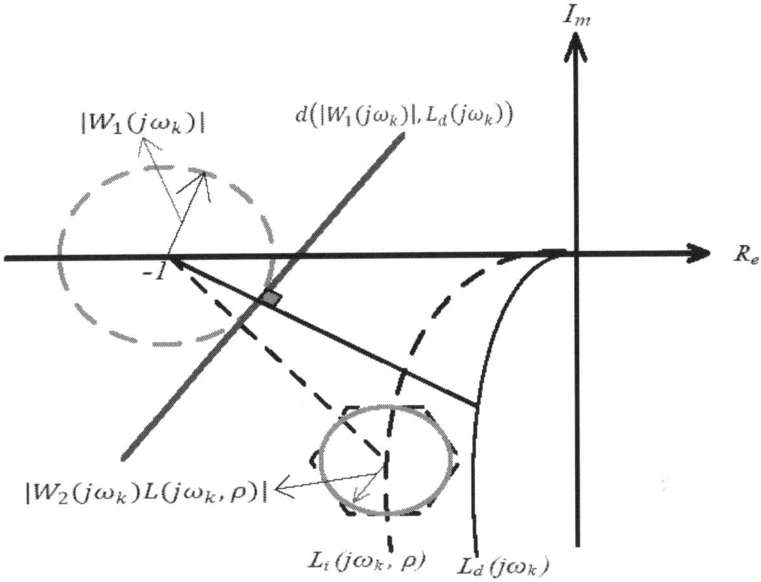

Figure 1. Linear constraints on Nyquist plot.

where $\sin(\alpha)$ and $\cos(\alpha)$ are functions of $L_d(j\omega_k)$, x and y are the real and imaginary parts on a point complex plane.

$$\sin(\alpha) = \frac{R_e\{1 + L_d(j\omega_k)\}}{|1 + L_d(j\omega_k)|}, \quad \cos(\alpha) = -\frac{I_m\{1 + L_d(j\omega_k)\}}{|1 + L_d(j\omega_k)|}$$

So the equation of the line d in (7) can be rewritten as:

$$|W_1(j\omega_k)[1 + L_d(j\omega_k)]| - I_m\{L_d(j\omega_k)\}y - [1 + R_e\{L_d(j\omega_k)\}][1 + x] = 0 \tag{8}$$

The side of the line d that excludes the critical point can be given by the following linear constraints:

$$\begin{aligned} &|W_1(j\omega_k)[1 + L_d(j\omega_k)]| - I_m\{L_d(j\omega_k)\}I_m\{L(j\omega_k,\rho)\} \\ &-[1 + R_e\{L_d(j\omega_k)\}][1 + R_e\{L(j\omega_k,\rho)\}] < 0 \ \forall\omega \end{aligned} \tag{9}$$

The linear constraints in (9) can be simplified by using the following facts:

$R_e\{L_d(j\omega_k)\}=1/2[L_d(j\omega_k)+L_d^*(j\omega_k)]$ and

$I_m\{L_d(j\omega_k)\}=1/2[L_d(j\omega_k)-L_d^*(j\omega_k)]$

The new linear constraints in (9) will be

$$|W_1(j\omega_k)[1+L_d(j\omega_k)]|-R_e\{[1+L_d^*(j\omega_k)][1+L(j\omega_k,\rho)]\}<0 \quad \forall\omega \text{ and } i=1,\cdots,v$$

(10)

where $L_d^*(j\omega_k)$ stands for the complex conjugate of $L_d(j\omega_k)$.

In order for this condition to be satisfied for a set of uncertainty models, the uncertainty circle centered at $L(j\omega_k,\rho)$ should be approximated by a polygon of $v>2$ vertices. Then, to satisfy the robust condition in (5) all the vertices of the polygon have to be located on the right side of the line d. This condition can be represented by the linear constraints on (11):

$$|W_1(j\omega_k)[1+L_d(j\omega_k)]|-R_e\{[1+L_d^*(j\omega_k)][1+L_i(j\omega_k,\rho)]\}<0 \quad \forall\omega \tag{11}$$

where $L_i(j\omega_k,\rho)=K(j\omega_k,\rho)G_i(j\omega)$, and

$$G_i(j\omega)=G(j\omega)\left[1+\frac{|W_2(j\omega_k)|}{\cos(\pi/v)}e^{-2j\pi i/v}\right] \tag{12}$$

Using all the above analysis, the quadratic optimization problem can be formulated as given in (13).

$$\min_{\rho}\sum_{i=1}^{m}\sum_{k=1}^{N_i}|L_i(j\omega_k,\rho)-L_d(j\omega_k)|^2 \tag{13}$$

Subject to:

$$\left|W_1(j\omega_k)\left[1+L_d(j\omega_k)\right]\right| - R_e\left\{\left[1+L_d^*(j\omega_k)\right]\left[1+L_i(j\omega_k,\rho)\right]\right\} < 0 \quad \forall\omega$$

for $k=1,\cdots,N_i$ (No. of frequencies), $i=1\cdots,m.$

where $L_i(j\omega_k,\rho)=\rho^{\mathrm{T}}\varphi(j\omega_k)G_i(j\omega_k).$

2.2. Choice of the Desired Open-Loop Transfer Function (L_d)

Selecting the desired open-loop transfer function should be based on considering plant, desired specifications and the structure of the designed controller. The desired open-loop transfer function L_d typically has a large amplitude in low frequencies (for good performance in tracking and disturbance rejection) and small amplitude in high frequencies for robustness with respect to unmodeled dynamics. A very typical choice for L_d is ω_c/s where ω_c the desired closed-loop bandwidth. It is clear that $|L_d|$ is large in low frequencies and small in high frequencies and crosses the 0 dB axis at the crossover frequency ω_c, which ensures a closed loop bandwidth equal to ω_c [7] [9].

2.3. The Controller Design Procedure

The steps of designing the proposed controller are:

1) Selecting different operating points. For case study, three operating points are given in Table 2.

2) Selecting a desired open-loop transfer function L_{d0}, typically equal to ω_c/s.

3) Solving the convex optimization problem in (13) to find an initial controller K_0.

4) The results may be improved if different desired open-loop transfer function is used for each model. In this case $L_{di} = K_0 G_i$ can be chosen.

5) The new desired open-loop transfer function L_{di} is used to design a new controller K for all the operating points.

3. TEST SYSTEM

3.1. Tow Area Four Machines Test System

The two-area four-machine test system is considered as one of the standard models that have been used to study the inter-area oscillations phenomenon. The test power system consists of two areas connected through two parallel tie lines and each area consists of two synchronous generators as shown in Figure 2. The generators of Area 1 (G1 and G2) are rated at 700 MW, and the generators of Area 2 (G3 and G4) are rated at 719 MW, and 700 MW, respectively. The four generators are equipped with automatic voltage regulators power system stabilizers, and turbine governors. The tie line power flow from Area 1 to Area 2 at the steady state is 400 MW. The SVC is installed at bus 8. The complete system data can be found in [10] .

3.2. Static VAR Compensator

SVC has been integrated in the test system. SVC is a shunt FACTS device that has been used to inject reactive current into the system at the point of connection. It is mainly used to improve voltage stability of the power system. SVCs can also be used for damping inter-area oscillations which can be achieved by adding a supplementary control signal to the voltage reference point of a SVC [5] . The location of SVCs in an interconnected power system might be at the sending end or receiving end of long tie-lines. The block diagramhj of the SVC that used in this paper is shown inFigure 3(a). The aim is to damp tie line oscillations using the SVC supplementary signal.

Table 1 shows the eigenvalue pair, the frequency and the damping ratio which represent the inter-area mode at the normal operating point $P_{tie} = 400\ \text{MW}$. The measured signal y is the tie-line power through the line 7 - 8 which is used as an input to the controller K(s) as shown in Figure 2. The output signal of the supplementary controller u is used to provide supplementary signal to the reference of the SVC. Therefore, the control system can be represented as shown in Figure 3(b). The parameters for the SVC are given in the Appendix.

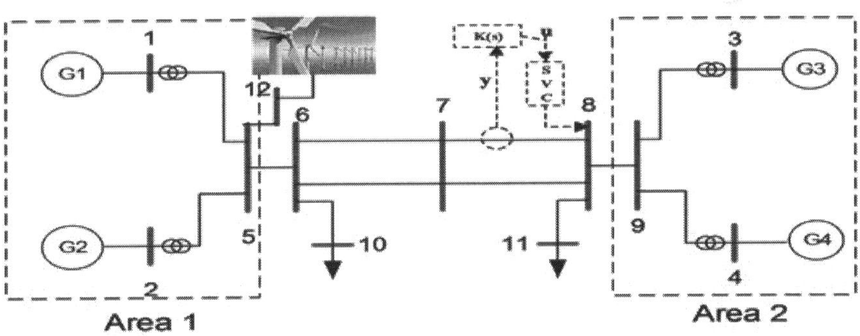

Figure 2. Single line diagram of two-area four-machines test system.

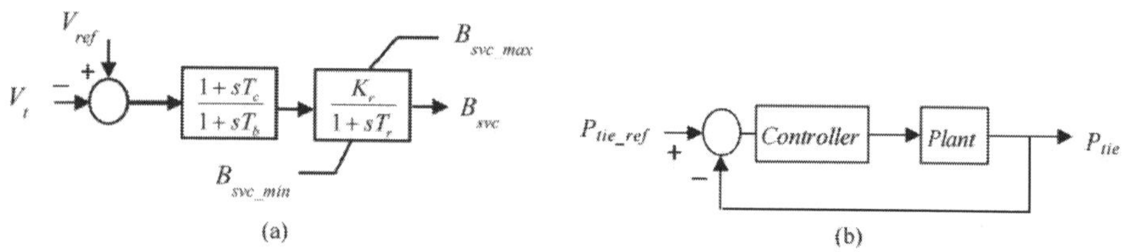

Figure 3. (a) The block diagram of an SVC; (b) The control representation of the system.

Table 1. Eigenvalue, damping ratio and the frequency of the test system.

Eigenvalue $\sigma \pm j\omega$	Damping ratio $-\frac{\sigma}{\sqrt{\sigma^2+\omega^2}}$	Frequency (Hz) $\frac{\omega}{2\pi}$
$-0.1114 \pm 4.044j$	0.02753	0.6435

3.3. DFIG Model

In this work, the third order model of a doubly fed induction generator is used. The dynamic modeling of the DFIG involves a set of differential algebraic equations [11] which has been built in PST [12] [13] . The model has been verified under large and small disturbances by comparing the time domain result with PSAT software. The wind turbine is installed in area 1 at bus 12 as shown in Figure 2.

4. SIMULATION RESULTS FOR THE CASE STUDY

4.1. Control Design

Three operating points (2, 4 and 5) as shown in Table 2 are selected to design the initial controller K_0 and the rest of the operating points are selected to test the robustness of the system. Both K_0 and L_{d0} are used to calculate the new open-loop transfer function L_{di} for all the models which is needed to design $K(s)$. The parameter of Laguerre basis is chosen as $\zeta = 1$ and the order of the controller is considered to be 3. The desired closed-loop bandwidth is $\omega_c = 10$ rad/sec, so the desired open loop transfer function is chosen (based on [9]) as $L_{do} = 10/s$. This choice is suitable for damping the desired inter-area modes. Tuned weighting filters W_1 as a low pass filter and W_2 as high pass filter are shown in (14). The transfer function of the designed third order controller is shown in (15).

$$W_1(s) = \frac{0.2s + 9.8}{s + 0.098}, \quad W_2(s) = \frac{0.04s^2 + 0.283s + 1}{1.6e - 5s^2 + 0.113s + 400} \tag{14}$$

$$K(s) = \frac{0.2152(s + 190.9)(s^2 + 0.92s + 2.29)}{(s+1)^3} \tag{15}$$

4.2. Two-Area Test System under Different Load Conditions

The two-area system is studied under different operating points (load conditions shown in Table 2) and fault conditions with and without the SVC supplementary controller $K(s)$. All the values inTable 2 are in per unit system based on 100 MVA. In this section the output power of the wind is kept constant at 300 MW.

4.2.1. The Controller Response to Different Load Conditions and Uncertainty in the System

To test the robustness of the test system, a three-phase to ground fault is applied at different locations (bus 6 and bus 8) and it is cleared (self-cleared) after 50 ms at different load conditions (points No. 1*, 4, and 6*) as shown in Table 2. As it has been mentioned the operating pint 4 is selected as one of the three operating points that the controller has been designed based on. On the other hand, the points 1* and 6* are selected as new load condition scenarios to see if the controller can maintain the robustness under these unconsidered points in the controller design. Figures 4(a)-(d) show the time domain results of the tie-line power in line 7 - 8 of the three operating points under three phase fault applied at bus 6 and bus 8. It can be seen that, the controller can maintain the robustness under different load conditions. Also, without the supplementary controller the oscillations need more time to be damped compared to the system with the controller.

4.2.2. The Controller Response to Changes in System Topology

The change in system topology has been achieved by applying three phase fault at bus 7 and clearing the fault by tripping the line 6 - 7. Figure 4(e) and Figure 4(f) show the tie-line power in line 7 - 8 at different load conditions under tripping the above line. The system cannot maintain the stability without the supplementary controller when the tie-line power between the two areas increased to be almost 500 MW as show in Figure 4(e); however, the stability can be maintained by adding the supplementary controller and that makes the system more robust. Adding the supplementary controller not only helps to damp the inter-area oscillations, it gives more flexibility to exchange the power between the two areas.

Table 2. Different operating points for two-area test system.

Operating Point No.	Load of Area 1	Load of Area 2	Approximate Tie-Line Power
1*	9.20	18.14	4.8
2	9.76	17.67	4
3*	10.76	16.75	3
4	11.76	15.77	2
5	12.66	14.82	1
6*	14.65	12.82	−1

*It is not used in the control design, it is used to validate the controller.

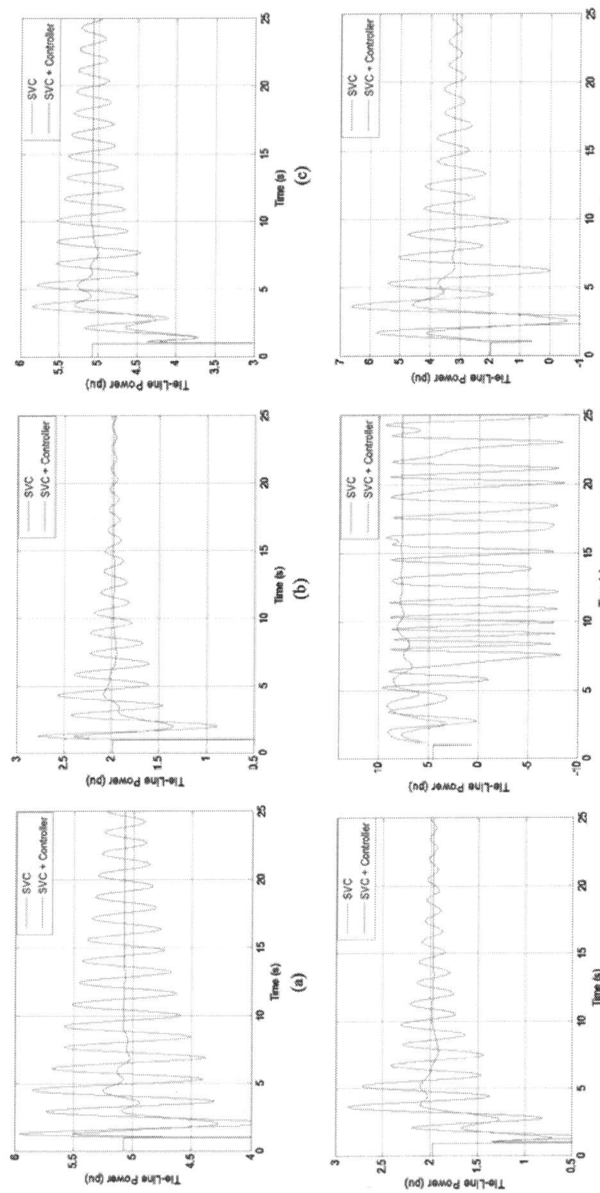

Figure 4. Tie-line power at different, load conditions, fault locations and change in system topology. (a) 480 MW tie-line, fault at bus 6; (b) 200 MW tie-line, fault at bus 6; (c) 480 MW tie line, fault at bus 8; (d) 200 MW tie line, fault at bus 8; (e) 480 MW tie-line, trip the line 6 - 7; (f) 200 MW tie-line, trip the line 6 - 7.

4.3. Two-Area Test System under Different Wind Penetrations

4.3.1. The Controller Response to Different Wind Penetrations and Uncertainty

The wind penetrations are changed according to the weather conditions and that affects the robustness of the system. The designed controller is tested under different levels of wind penetrations. Three different levels of wind penetrations (300, 200 and 100 MW) are selected to show the impact of wind power penetration on the robustness of the test system. Figures 5(a)-(c) show the tie-line power under different operating points when the level of the wind penetration is equal to 200 MW and 100 MW. The simulation results illustrate that, the supplementary controller is able to damp the oscillations within 5 sec under different load conditions.

4.3.2. The Controller Response to Changes in System Topology

The performance of the designed controller has been examined under tripping the line 7 - 8 at different level of wind penetrations. Figures 5(d)-(f) show the tie-line power in line with 200 MW and 100 MW wind under different load conditions. The controller can maintain the stability under this outage which makes the system more robust.

4.4. Eigenvalue Analysis

Eigenvalue study has been done to examine the performance of the supplementary controller in terms of improving the damping ratio ξ of the inter-area modes. The results are concluded in Table 3. It can be seen that the damping ratios of different load conditions are improved.

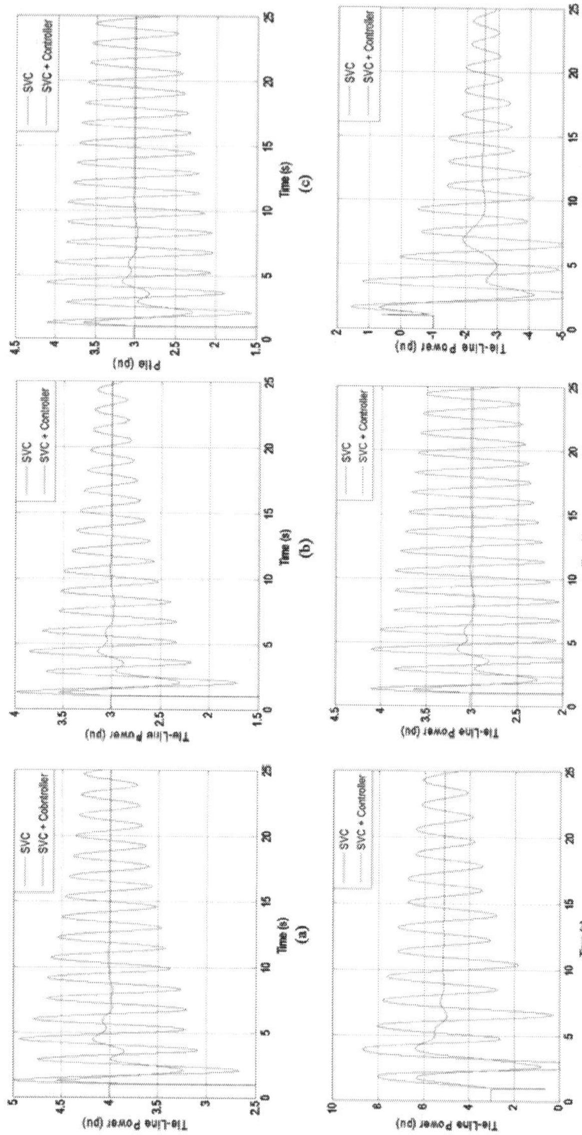

Figure 5. Tie-line power at different wind penetrations and change in system topology. (a) 400 MW, 200 MW wind, fault at bus 6; (b) 300 MW, 200 MW wind, fault at bus 6; (c) 300 MW, 100 MW wind, fault at bus 6; (d) 300 MW, 200 MW wind, trip the line 6 - 7; (e) 300 MW, 100 MW wind, trip the line 6 - 7; (f) 100 MW, 100 MW wind, trip the line 6 - 7.

Table 3. Damping and frequencies of the inter-area modes under different load conditions.

Tie-line power (pu)	SVC		SVC with controller	
	ξ	f(Hz)	ξ	f(Hz)
5	0.0247	0.6386	0.2908	0.7319
3	0.0332	0.6505	0.4005	0.6794
2	0.0385	0.6550	0.4537	0.6295
1	0.0432	0.6567	0.4882	0.5845

Table 4. Damping and frequencies under different wind penetrations.

Level of wind penetrations (pu)	SVC		SVC with controller	
	ξ	f(Hz)	ξ	f(Hz)
3	0.0275	0.6434	0.3311	0.7195
2	0.0139	0.6425	0.3280	0.7231
1	0.0025	0.6408	0.3263	0.7228
0.5	−0.0009	0.6388	0.3273	0.7225

Table 4 summarized the damping ratios of the inter-area modes under different level of wind penetrations ($P_{tie} = 400\text{ MW}$). The results show that the action of the supplementary controller is robust against varying the level of the wind penetrations. It can be seen that when the penetration level is decreased to 50 MW the system becomes unstable (negative damping ratio) without the controller.

5. CONCLUSION

A new method was introduced and implemented in this paper to design a robust fixed-order H_∞ controller. The basic concept of the H_∞ design has been detailed. Basically this approach is based on shaping the open-loop transfer function in the Nyquist diagram. The performance of the designed controller has been evaluated under different load conditions as well as different levels of wind penetrations. One of the advantages of the proposed approach is that it does not need the system to be reduced. For example, the two-area test system with wind has 75 states and only 3-order controller is required. The main advantage of the proposed method is considering the multi-model uncertainty. The results show that the designed controller is robust and helps the system to maintain the stability under small and large disturbances. In some cases, the system loses the stability without the supplementary controller. It can be concluded that the controller is robust respect to changes in the load conditions as well as different levels of wind penetrations.

REFERENCES

1. Chaudhuri, B. and Pal, B. (2004) Robust Damping of Multiple Swing Modes Employing Global Stabilizing Signals with a TCSC. IEEE Transactions on Power Systems, 19, 499-506.

2. Chaudhuri, B., Pal, B., Zolotas, A.C., Jaimoukha, I.M. and Green, T.C. (2003) Mixed-Sensitivity Approach to H Control of Power System Oscillations Employing Multiple Facts Devices. IEEE Transactions on Power Systems, 18, 1149- 1156.

3. Klein, M., Le, L., Rogers, G., Farrokpay, S. and Balu, N. (1995) H Damping Controller Design in Large Power System. IEEE Transactions on Power Systems, 10, 158-166.

4. Majumder, R., Pal, B.C., Dufour, C. and Korba, P. (2006) Design and Real-Time Implementation of Robust FACTS Controller for Damping Inter-Area Oscillation. IEEE Transactions on Power Systems, 21, 809-816.

5. Zhao, Q. and Jiang, J. (1995) Robust SVC Controller Design for Improving Power System Damping. IEEE Transactions on Power Systems, 10, 1927-1932.

6. Chen, S. and Malik, O. (1995) Power System Stabilizer Design Using Synthesis. IEEE Transactions on Energy Conversions, 10, 175-181.

7. Karimi, A. (2013) Frequency-Domain Robust Control Toolbox. 52nd Conference in Decision and Control, Florence, Italy, 10-13 December 2013, 3744-3749.

8. Galdos, G. (2010) Fixed-Order Robust Controller Design by Convex Optimization Using Spectral Models. Thèse EPFL, No. 4785.

9. Karimi, A. and Galdos, G. (2010) Fixed-Order H∞ Controller Design for Nonparametric Models by Convex Optimization. Automatica, 46, 1388-1394.

10. Kundur, P. (1994) Power System Stability and Control. McGraw-Hill, New York.

11. Milano, F. (2005) Power System Analysis Toolbox.

12. MathWorks Inc. (2014) MATLAB Version 8.3.0.532.

13. Cherry Tree Scientific Software (1991-2008) Power System Toobox ver. 3.0.

Chapter 3

Monitoring Power Quality in Small Scale Renewable Energy Sources Supplying Distribution Systems

Nicolae Golovanov[1], George Cristian Lazaroiu[1], Mariacristina Roscia[2] and Dario Zaninelli[3]

[1] Department of Power Systems, University Politehnica of Bucharest, Romania

[2] Department of Design and Technology, University of Bergamo, Italy

[3] Dipartimento di Energia, Politecnico di Milano, Italia

1. INTRODUCTION

The integration of renewable sources within the existing power system affects its traditional principles of operation. The renewable energy sources (RES) can be used in small, decentralized power plants or in large ones, they can be built in small capacities and can be used in different locations [1]. In isolated areas where the cost of the extension of the power systems (from utilities point of view) or the cost for interconnection with the grid (from customer's point of view) are very high with respect to the cost of the RES system, these renewable sources are suitable. The RES systems are appropriate for a large series of applications, such as stand-alone systems for isolated buildings or large interconnected networks. The modularity of these systems makes possible the extension in the case of a load growth.

The increasing penetration rate of RES in the power systems is raising technical problems, as voltage regulation, network protection coordination, loss of mains detection, and RES operation following disturbances on the distribution network [2]. The utilization of these alternative sources presents advantages and disadvantages. The impact of the wind turbines and photovoltaic systems on network operation and power quality (harmonics, and voltage fluctuations) is highly important. The capability of the power system to absorb the power quality disturbances is depending on the fault level at the point of common coupling. [3] In weak networks or in power systems with a high wind generation penetration, the integration of these sources can be limited by the flicker level that must not exceed the standardized limits. The wind generators and PV systems interconnected to the main grid with the help of power electronics converters can cause important current harmonics.

2. POWER QUALITY AND RENEWABLE ENERGY SOURCES

Nowadays, the renewable sources generation is rapidly developing in Europe. In the last 17th years, the average growth rate is of wind generation is 15.6% annually [4]. As these renewable sources are increasingly penetrating the power systems, the impact of the RES on network operation and power quality is becoming important. The intermittent character of the wind and solar irradiation constrains the power system to have an available power reserve. Due to the output power variations of the wind turbines, voltage fluctuations are produced. In weak networks or in power systems with a high wind generation penetration, the integration of these sources can be limited by the flicker level that must not exceed the standardized limits.

The photovoltaic (PV) installations, interconnected to the mains supply, can be single-phase connected (photovoltaic installations with capacity less than 5 kW) or three-phase connected (photovoltaic installations with capacity greater than 5 kW). The direct-coupled PV systems, without electrical energy storage, inject in the power system a generated power that follows the intermittency of the primary energy source. In this case, important voltage variations can occur at the PCC. The connection of PV systems to the low voltage grid can determine voltage variations and harmonic currents [5].

2.1. Voltage Fluctuations

Determination of voltage fluctuations (flicker effect) due to output power variations of renewable sources is difficult, because depend of the source's type, of generator's characteristics and network impedance. For the case of wind turbines, the long term flicker coefficient *Plt* due to commutations, computed over a 120 min interval and for step variations, becomes [6]:

$$P_{lt} = \frac{8}{S_{sc}} \cdot N_{120}^{0.31} \cdot k_f\left(\psi_{sc}\right) \cdot S_r \qquad (1)$$

where *N120* is the number of possible commutations in a 120 min interval, *kf*(ψ*sc*) is the flicker factor defined for angle ψ_{sc} = arctan*(Xsc /Rsc)*, *Sr* – rated power of the installation, and *Ssc* – fault level at point of common coupling (PCC).

For a 10 minutes interval, the short-term flicker *Pst* is defined [6]:

$$P_{lt} = \frac{18}{S_{sc}} \cdot N_{10}^{0.31} \cdot k_f\left(\psi_{sc}\right) \cdot S_r \qquad (2)$$

where *N10* is the number of possible commutations in a 10 min interval.

The values of flicker indicator for wind turbines, due to normal operation, can be evaluated using flicker coefficient $c(\psi_{sc}, \upsilon_a)$, dependent on average annual wind speed, υ_a, in the point where the wind turbine is installed, and the phase angle of short circuit impedance, ψ_{sc}:

$$P_{st} = P_{lt} = c\left(\psi_{sc}, v_a\right)\frac{S_r}{S_{sc}}. \qquad (3)$$

The flicker coefficient $c(\psi_{sc}, \upsilon_a)$ for a specified value of the angle ψ_{sc}, for a specified value of the wind speed υ_a and for a certain installation is given by the installation manufacturer, or can be experimentally determined based on standard procedures. Depending on the voltage level where the wind generator (wind farms) is connected, the angle ψ_{sc} can take values between 30° (for the medium voltage network) and 85° (for the high voltage network). Flicker evaluation is based on the IEC standard 61000-3-7 [7] which gives guidelines for emission limits for fluctuating loads in medium voltage and high voltage networks. Table 1 reports the recommended values.

The flicker evaluation determined by a wind turbine of 650kW is analyzed. The wind turbine has a tower height of 80 meters, the rotor diameter is 47 m, and the swept area is 1735 m^2. The electrical energy production during two winter months is 127095 kWh, respectively 192782 kWh. The average wind speeds, measured at 60m height, during the first monitoring month was 6.37m/s while during march was 7.32m/s. The variation of turbine output power is shown in Fig. 1. The intermittent character of the produced power is clearly highlighted. The tower shadow effect for the wind generator determines a variation of the absorbed energy, which is measured as a power variation at generator terminals. Fig. 2(a) shows the wind generator, while Fig. 2(b) illustrates the tower shadow effect corresponding variation of the generator output power.

Table 1. Flicker planning levels for medium voltage (MV) and high voltage (HV) networks

Flicker severity factor	Planning levels	
	MV	HV
P_{st}	0.9	0.8
P_{lt}	0.7	0.6

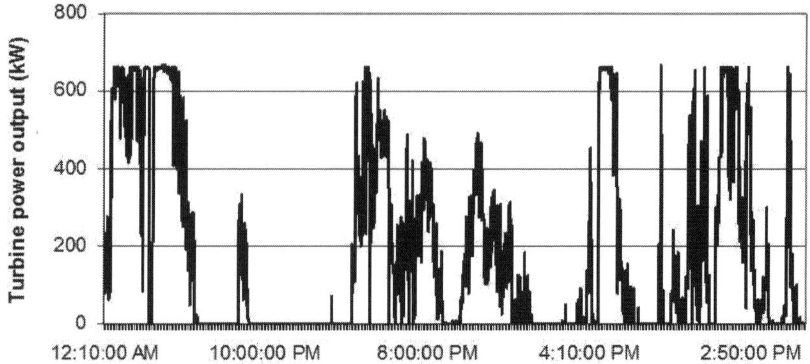

Figure 1. Turbine output power variation during the 1 month monitoring period

The measured values of the flicker coefficient $c(\psi_{sc}, \upsilon a)$ for different values of the annual average wind speed υa and for different network impedance angle ψ_{sc} are reported in Table 2. Table 3 reports the flicker coefficient *kf* values for voltage step variations, for the same wind generator.

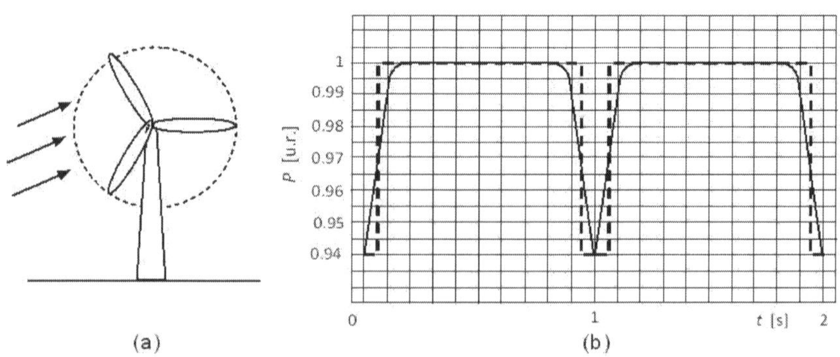

Figure 2. Variation of the wind generator output power due to the tower shadow effect.

Table 2. Values of the flicker factor for various values of the wind speed *va* and for various angles ψ_{sc}

Annual wind speed v_a [m/s]	Network impedance angle ψ_{sc} [°]			
30°	50°	70°	85°	
6	3.1	2.9	3.6	4.0
7.5	3.1	3.0	3.8	4.2
8.5	3.1	3.0	3.8	4.2
10	3.1	3.1	3.8	4.2

Table 3. Values of the flicker factor *kf*

		Network impedance angle ψ_{sc} [°]			
		30°	50°	70°	85°
Flicker factor k_f for voltage step variations	With start at minimum speed	0.02	0.02	0.01	0.01
	With start at rated speed	0.12	0.09	0.06	0.06
Installation is sized for $N_{10} = 3$; $N_{120}=35$					

The computations based on the values reported in Table 2 and Table 3 lead to the flicker indicator values:

1) continuous operation, annual average wind speed v_a=7.5, interconnection with the medium voltage network (ψ_{sc}=50°, S_{sc} = 300 MVA, S_r = 0.65MVA)

$$P_{st} = P_{lt} = \frac{S_r}{S_{sc}} \cdot c(\psi_{sc}, v_a) = \frac{0.65}{300} \cdot 3 = 0.0065$$

2. generator interconnection at minimum speed of the wind turbine

$$P_{st} = 18 \cdot N_{10}^{0,31} \cdot k_f(\psi_{sc}) \cdot \frac{S_r}{S_{sc}} = 18 \cdot 3^{0,31} \cdot 0,02 \cdot \frac{0.65}{300} = 0,00109;$$

$$P_{lt} = 8 \cdot N_{120}^{0,31} \cdot k_f(\psi_{sc}) \cdot \frac{S_r}{S_{sc}} = 8 \cdot 35^{0,31} \cdot 0,02 \cdot \frac{0.65}{300} = 0,00104.$$

3. generator interconnection at rated speed of the wind turbine

$$P_{st} = 18 \cdot N_{10}^{0,31} \cdot k_f(\psi_{sc}) \cdot \frac{S_r}{S_{sc}} = 18 \cdot 3^{0,31} \cdot 0,09 \cdot \frac{0.65}{300} = 0,0049;$$

$$P_{lt} = 8 \cdot N_{120}^{0,31} \cdot k_f(\psi_{sc}) \cdot \frac{S_r}{S_{sc}} = 8 \cdot 35^{0,31} \cdot 0,09 \cdot \frac{0.65}{300} = 0,0047.$$

Due to the output power variations of the wind turbines, voltage fluctuations are produced. Voltage fluctuations are produced due to the wind turbine switching operations (start or stop), and due to the continuous operation. The presented voltage fluctuations study, made for one turbine, becomes necessary in large wind farms as the wind power penetration level increases quickly.

2.2. Pv Impact On Steady State Voltage Variations

The variability nature of solar radiation, the weather changes or passing clouds can cause important variation of PV output power [8]. The variation of the power produced by a 30kW PV system is illustrated in Fig. 3. Fig. 4 (a) shows the generation power in a sunny day, while Fig. 4 (b) illustrates the generation power in a cloudy day.

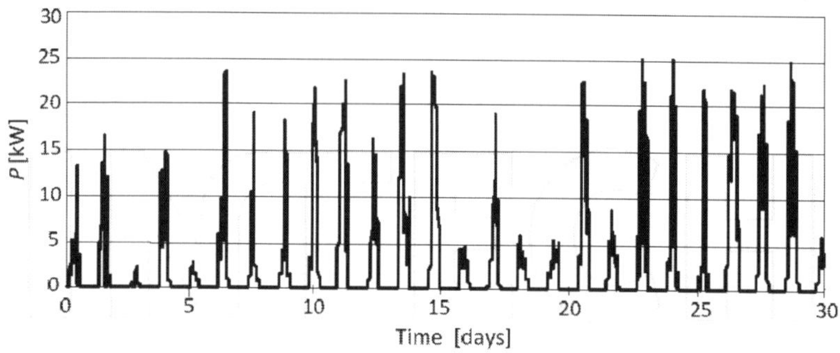

Figure 3. Variation of PV system power output during a month

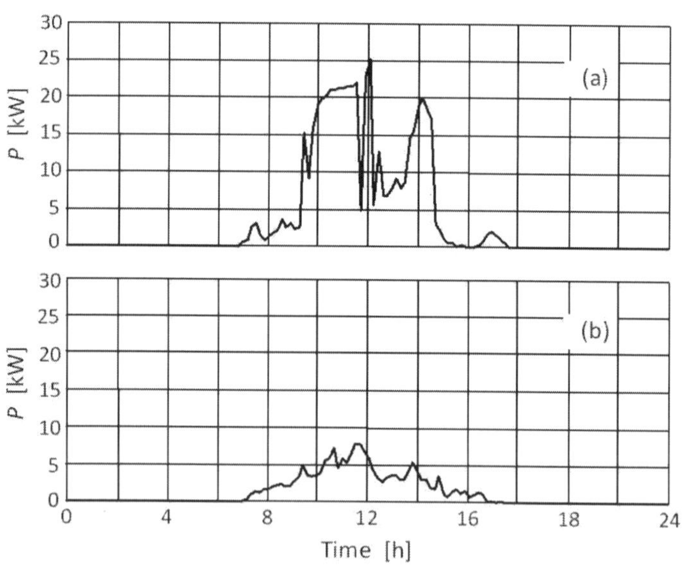

Figure 4. Variation of PV system power output during: (a) sunny day, (b) cloudy day

The connection of these variable renewable sources can determine a voltage rise at PCC and in the grid. The utility has the general obligation to ensure that customer voltages are kept within prescribed limits. A voltage variation ΔV between *Vmax* and *Vmin*, can appear on short periods. This voltage variation can highly stress the electrical devices supplied by the power system, and in particular the owner of the photovoltaic facility (as shown in Fig. 5). Fig. 5 illustrates the possible case of a summer mid-day, when the load downstream PCC is relatively small and the PV output power exceeds the demand.

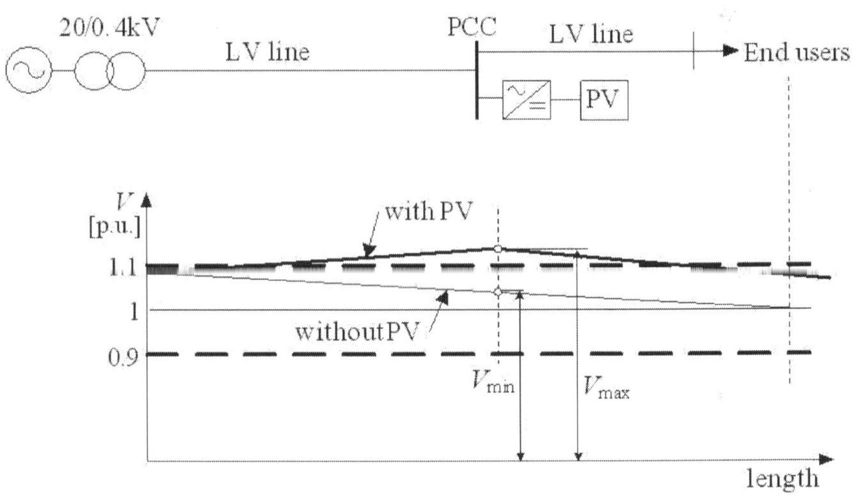

Figure 5. Influence of PV on voltage level

Voltage variations can influence the characteristics of the electrical equipment and household appliances (loss of the guaranteed performances, modifications of the efficiency) leading in some cases even to the interruption of operation. The voltage variation at PCC can be expressed as:

$$\Delta V = \frac{S_{PV}}{S_{sc}} \cdot \cos(\psi_{sc} - \varphi) \tag{4}$$

where *SPV* is the power produced by PV, S_{sc} is the short circuit power at PCC, ψ_{sc} = arctan(X/R) is the angle of the network short circuit impedance, ϕ is the phase angle of the PV output current (we consider that the electric quantities are sinusoidal). In existing power systems, there are measures such that the line voltage to be sinusoidal. In (4), the system harmonics are not considered.

At low and medium voltage levels, the utilities have established limits for the amplitude of the voltage variations, which must not be exceeded during normal operation. Due to the statistical nature of the steady state voltage variations, the standard EN 50160 stipulates statistical limits [9]. In some Countries the limit ±10% established in EN 50160 is applied, while other present guidelines, elaborated in different Countries, impose more restrictive limits for the voltage variations. The relevant variations of the voltage will overlap the voltage's variations caused by load modification and can lead to the widening of the voltage limit bands.

2.3. Current Harmonic Perturbations

Measurement results of a 200 MW wind farm reveals the harmonic current and voltage spectra, active and reactive power variations, and the relationship between wind farm harmonic emission level and output power. It is considered that the harmonics are determined by the converter at the interconnection point with the main power system. In order to connect the PV power systems with the grid, an inverter that transforms the dc output power of the PV to the 50Hz ac power is required. The small capacity PV systems are interconnected to the main grid with the help of simple single-phase inverters, which can cause important current harmonics. General requirements can be found in standards, especially those for the interconnection of distributed generation systems to the grid and for photovoltaic systems [1, 10]. In the standard IEEE 1547, the harmonic current injection of RES at the PCC must not exceed the limits stated in table 4.

Table 4. Maximum harmonic current distortion in percent of current *I*, where *I* is the fundamental frequency current at full system output. Even harmonics in these ranges shall be <25% of the odd harmonic limits listed [**1**].

Individual harmonic order	h<11	11≤h<17	17≤h<23	23≤h<35	35≤h	TDD
Percent (%)	4	2	1.5	0.6	0.3	5

The current variation on phase *a*, during one week monitoring period of the 200 MW wind farm, is illustrated in Fig. 6. The high current variability function of wind speed can be clearly observed. The variation of total current harmonic distortion (*THDI*) during the monitoring period is illustrated in Fig. 7. The inverse relationship between the RMS electrical current and *THDI* can be seen in Fig. 6 andFig. 7, at the same time instants. When the generated power is high (large value of RMS electrical current), the fundamental current is high and the *THDI* is small. For small generated powers, the fundamental current is small and the *THDI* is high. From practical point of view, this fact is not highly important as the small current values do not influence the voltage quality at point of common coupling.

Fig. 8 illustrates the output power variation of a PV system and the total current harmonic distortion variation during a day [11]. When the PV system has an output power close to the rated power, the*THDI* is relatively low. During the shadowing period of the PV system, the *THDI* is taking high values.

The analysis of the total harmonic distortion factor has to consider that, for high variability of the primary energy source, the large *THDI* values can lead to inappropriate conclusions. For the periods with small primary energy source, the electric current injected into the grid presents a reduced fundamental component, resulting in a high distortion factor. As the electrical current has small values, the voltage distortion and the voltage drop in the power system are negligible, and thus the voltage waveform at the point of common coupling is not affected.

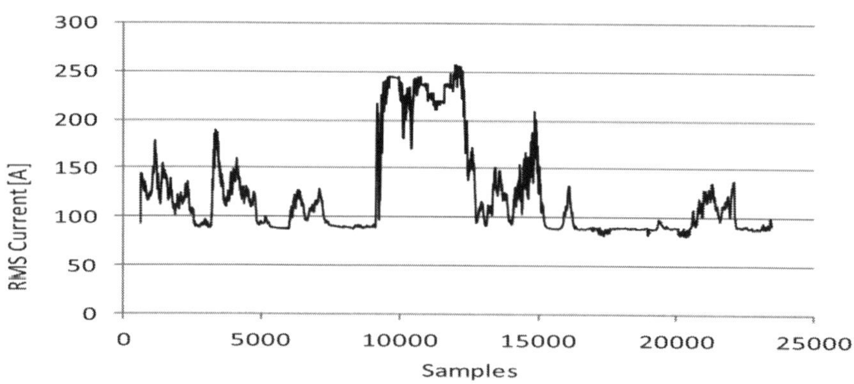

Figure 6. Variation of RMS current, phase a, during the monitoring period

3. CONCLUSIONS

The renewable sources interconnected with the main supply can influence the power quality at the point of common coupling and can pollute the electrical network with harmonic components that must not exceed the stipulated limits. The existing trend of installing more and more small capacity sources implies the establishment as accurate as possible of their impact on power system operation.

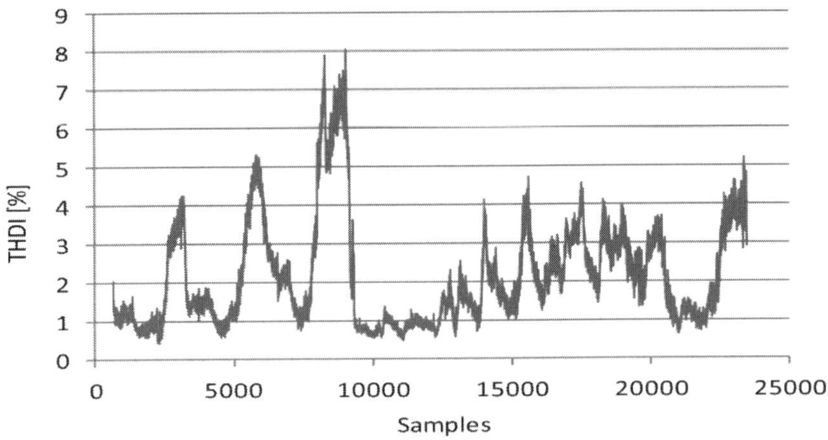

Figure 7. Variation of *THDI* during the monitoring period

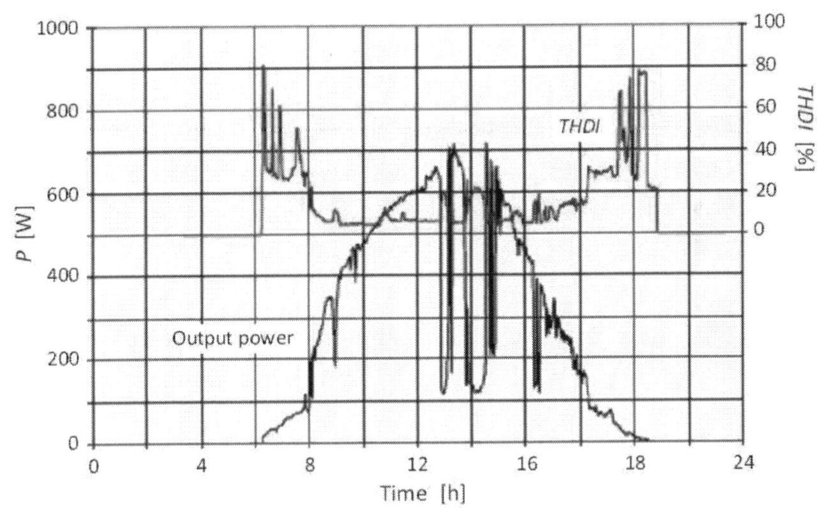

Figure 8. Variation of PV system output power and variation of THDI, during a day.

The voltage fluctuations determined wind power variations are analyzed, both for the wind turbine switching operations (start or stop), as well as for the continuous operation. The voltage flicker study becomes necessary as the wind power penetration level increases quickly. The connection of variable renewable sources, like photovoltaic systems, can determine a voltage rise at PCC and in the grid which can affect the electrical characteristics of the equipments.

The wind generators and photovoltaic sources, connected to the power system through power electronic converters, can pollute the electrical network with harmonic components that must not exceed the stipulated limits. A better characterization, from the practical point of view, of the total current harmonic distortion determined by renewable energy sources interconnected to the mains supply through power electronic converters is necessary.

REFERENCES

1. IEEE Standard for interconnecting distributed resources with electric power systems, IEEE Standard 1547, 2003.

2. N. Jenkins and R. Allan, Embedded generation, The Institution of Electrical Engineers, London, 2000, pp. 49-85.

3. R. C. Dugan, H. W. Beaty, and M. F. McGranagham, Electrical power systems quality, New York: Mc.Graw-Hill, 1996, pp. 1-8.

4. European Wind Energy Association [Online]. Available: http://www.ewea.org/fileadmin/files/library/publications/statistics/Wind_in_power_2011_European_statistics.pdf

5. F. Gagliardi, F. Iannone, G.C. Lazaroiu, M. Roscia, D. Zaninelli, Sustainable building for the quality of power supply, in Proc. IEEE PowerTech Conference 2005, Sankt Petersburg, Russia, June 27-30, 2005, pp. 6

6. IEC 61400-21, Wind turbine generator systems - Measurement and assessment of power quality characteristics of grid connected wind turbines, 2001

7. IEC 61000-3-7, Assessment of emission limits for the connection of fluctuating load installation to MV, HV and EHV power systems, 2008.

8. M. Simonov, M. Mussetta, F. Grimaccia, S. Leva, R. E. Zich, "Artificial Intelligence forecast of PV plant production for integration in smart energy systems," International Review of Electrical Engineering, vol. 7, no. 1, pp. 3454-3460

9. EN 50160/2010, Voltage characteristics of electricity supplied by public distribution systems.

10. IEEE Recommended Practice for Utility Interface of Photovoltaic (PV) Systems, IEEE Standard 929-2000, April 2000

11. G. Chicco, J. Schlabbach, F. Spertino, Characterisation and Assessment of the Harmonic Emission of Grid-Connected Photovoltaic Systems, in Proc. IEEE Power Tech 2005, St. Petersburg, Russia, June 27-30, 2005, pp.7

Chapter 4

Hilbert-Huang transform with adaptive waveform matching extension and its application in power quality disturbance detection for microgrid

Peng LI[1] , Jing GAO[1] , Duo XU[1] , Chang WANG[1] , Xavier YANG[2]

[1]State Key Laboratory of Alternate Electrical Power System with Renewable Energy Sources, North China Electric Power University, Baoding 071003, China

[2] R&D of Electricite de France (EDF), China Center, Beijing, China

ABSTRACT

With the significant improvement of microgrid technology, microgrid has gained large-scale application. However, the existence of intermittent distributed generations, nonlinear loads and various electrical and electronic devices causes power quality problem in microgrid, especially in islanding mode. An accurate and fast disturbance detection method which is the premise of power quality control is necessary. Aiming at the end effect and the mode mixing of original Hilbert-Huang transform (HHT), an improved HHT with adaptive waveform matching extension is proposed in this paper. The innovative waveform matching extension method considers not only the depth of waveform, but also the rise time and fall time. Both simulations and field experiments have verified the correctness and validity of the improved HHT for power quality disturbance detection in microgrid.

KEYWORDS

Adaptive waveform matching extension, End effect, Improved Hilbert-Huang transform, Microgrid, Power quality

1. INTRODUCTION

Microgrid technology has provided a new technical approach for the large-scale integration of renewable energy and distributed generations, as well as technical support for the grid-connected operation of distributed generations to meet the requirements of smart grid. However, power quality problem is a prominent challenge for microgrid. On one hand, there are a variety of distributed generations and nonlinear fluctuating loads in microgrid. Especially the output power of wind turbines and photovoltaic cells has the trait of fluctuation, randomness and intermittent. This may lead to the unbalanced power between microsources and loads. On the other hand, the extensive use of power electronic devices such as grid-connected inverters, solid state switches and electric vehicle charging devices deteriorates the power quality indices. The research on power quality in microgrid has important theoretical and practical significance.

The power quality issue in microgrid may be dealt with a new power management system, accurate and rapid disturbance detection is an essential function of this system. Power quality issues in microgrid mainly include voltage and current harmonics, voltage sags, voltage swells, voltage short interruptions, voltage fluctuations and flickers, voltage and current unbalance components and so on [1–4]. Due to the traits of distributed sources the harmonics and inter-harmonics in microgrid may be abundant. Meanwhile the voltage fluctuation and flicker as random, dynamic and non-stationary phenomenon may become important issue for microgrid operation.

The disturbance detection method for microgrid application needs to analyze not only harmonic and inter-harmonic signals but also nonlinear and non-stationary signals. Hilbert-Huang transform (HHT) is an adaptive time-frequency analysis method which can deal with nonlinear and non-stationary signal analysis as well. This time-frequency analysis method can adaptively decompose signals according to their characteristics, thus characteristics of power quality disturbance are automatically extracted from the signals themselves. Compared with Fourier transform, HHT can analyze non-stationary and non-periodic signals. Furthermore, compared with wavelet transform, HHT not only has the advantages of wavelet transform, but also does not need to select basic functions [5]. For these reasons, HHT is a suitable method to carry out disturbance detection in microgrid. In this paper an improved HHT method is studied.

2. BASIC PRINCIPLE AND EXISTING PROBLEM OF HHT

2.1 Basic principle

HHT is a new nonlinear and non-stationary data analysis method composed of empirical mode decomposition (EMD) and Hilbert transform (HT). A nonlinear and non-stationary sequence can be decomposed into finite number of intrinsic mode functions (IMFs) and a trend term through EMD, which is called EMD sifting processing [6, 7]. An IMF must satisfy the following two constrains: ① the number of zero-crossing points and the number of extreme points must either be equal or differ at most by one; ② the average value of the upper envelope defined by the local maximums and the lower envelope defined by the local minimums is zero. Hilbert transform is applied to each IMF to calculate its instantaneous frequency and amplitude and to get the oscillation characteristics of different time scales which are contained in signals.

2.2 End effect

One of the greatest disadvantages of HHT is the end effect which exists in the process of EMD and HT. Signal distortion superimposes in the two processes so that the end portions of HHT cannot correctly reflect the information contained in the signal. Specifically, cubic spline interpolation function is used to obtain the upper and lower envelopes in EMD process. But it cannot ensure the two endpoints of the data sequence are exactly the extreme points. That means, at the endpoints, the spline interpolation may have low precision and it may lead to 'overshoot' or 'undershoot' phenomenon. Even worse, the entire data sequence may be "contaminated" through loop iteration, which eventually leads to serious distortion of IMFs and the generation of false IMFs.

Fast Fourier transform is exploited to compute instantaneous frequency and amplitude of IMFs during Hilbert transform. However if the periodic signal is sampled non-periodically, Fourier transform may cause 'Gibbs phenomenon' and frequency leakage, which may result in 'runaway' at both ends of the signal, that is, losing information of the signal itself.

The end effect reduces the signal analysis accuracy of HHT and impedes its wide application. Many methods and techniques have been studied and developed to solve this problem. Ideally, periodic sampling is the most valid way but its feasibility is poor. The essence of the suppression of the end effect in EMD is how to solve the interpolation of bad points in the process of fitting extreme point envelope curve. There are two basic methods for this problem: ① improving the interpolation function; ② extending the signal at the two endpoints to reduce the errors of fitting envelope curve at the endpoints. The latter method is the main solution for the end effect. It can be divided into three categories which are summarized and shown in Fig. 1 [8–13].

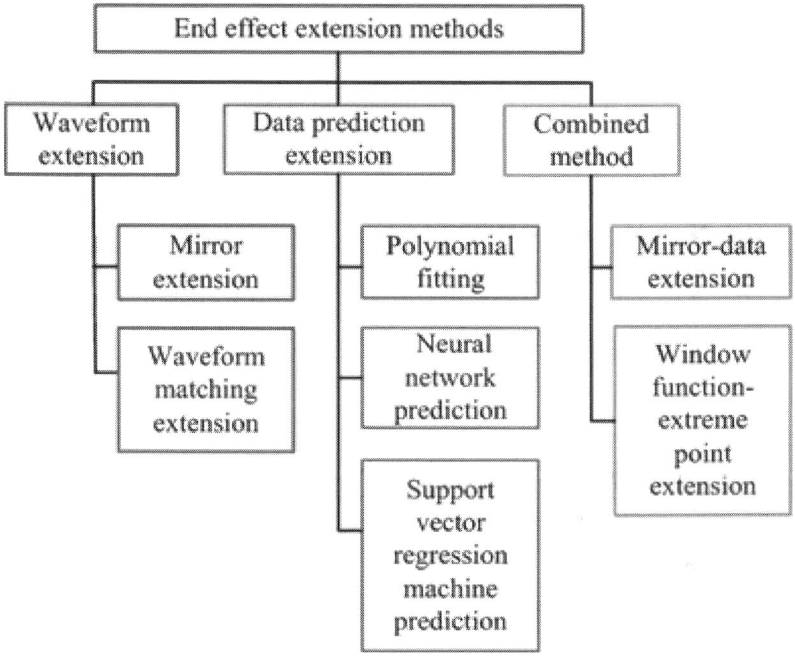

Figure 1. Extension methods for end effect prediction

Among them, the basic idea of waveform extension is to use the variation trend of the signal to extend itself. The extension results approach to the characteristics of the original data. The basic idea of data prediction extension is to build a particular mathematical model, of which the parameters are computed according to the original signal. The trend of the two endpoints of the signal can be predicted by the model. However, the establishment of the model (especially those of some artificial intelligence algorithms) is of high requirement. As for extreme point extension, it uses the extreme points to extend the signal. But its time adaptability is poor. In order to take full advantage of each method, some of these methods are combined. But the difficulty is how to find an effective way to make the combination.

2.3 Mode mixing

Another important drawback of HHT is the mode mixing which is discovered by Huang firstly. The mode mixing refers to that an IMF contains quite different characteristic time scales, or different IMFs contain similar characteristic time scales. It appears when the waveform of an IMF mixes with those of its adjacent IMFs. Thus identification of the waveforms is difficult which causes Hilbert transform to lose physical meaning [14]. There are two reasons of the generation of mode mixing. Firstly, signals are mixed with abnormal events [6], that is, discontinuous component contained in signals [7], pulse interference and noise. EMD uses the upper and lower envelopes of the signal (which are obtained by the spline interpolation of the extreme points of the signal) to compute the average value, then to complete the screening process. Therefore the presence of abnormal events leads to locally abnormal distribution of the extreme points, which results in mode mixing. Secondly, the amplitude and frequency of each component of the signal may influence each other. Because the amplitude and frequency determine the distribution of the extreme points, the influence between each component finally causes mode mixing.

Currently, solutions for mode mixing are proposed by researchers according to different causes. Figure 2 shows the classification of these methods [14–17]. For mode mixing caused by abnormal events, there are methods include abnormal event elimination method, auxiliary signal addition method, signal filtering method and so on. The adaptability of abnormal event elimination method needs further improvement. Signal filtering method cannot completely eliminate the abnormal events. For mode mixing caused by the influence of the amplitude and frequency of the signal itself, there are methods included pseudo wave method of additive and subtractive sinusoidal signal, overall average EMD added white noise and so on. But both the two methods have a problem that the physical meaning of the average IMF is controversial.

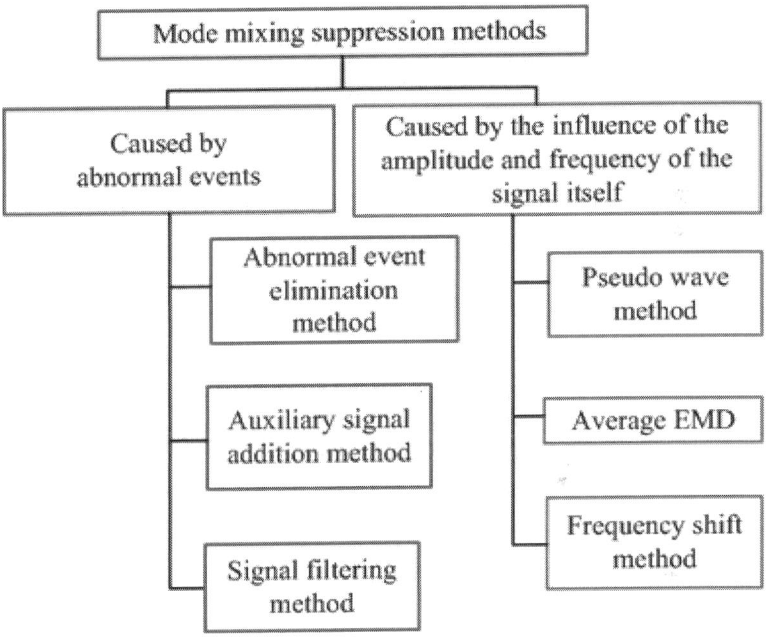

Figure 2. Suppression methods for mode mixing

3. IMPROVED HHT WITH ADAPTIVE WAVEFORM MATCHING EXTENSION

3.1 Improved HHT method

In order to improve the accuracy of HHT for power quality disturbance detection, this paper introduces an adaptive waveform matching extension method.

To solve the end effect of EMD, both ends of the original signal require extension. The extension should be consistent with the changing trend of the original signal as much as possible. So the changing trend of the original signal can be maintained and a smooth transition from the original signal to the extension waveform can be obtained to suppress the end effect more effectively. Based on this, the adaptive waveform matching extension method is proposed.

The basic idea of the adaptive waveform matching extension method is as follows. Based on the hypothesis that similar waveforms would be repeated in the signal, one can find a triangular waveform which matches the boundary of the signal best in the waveform of the signal. According to the local value of the triangular waveform, one can predict the local value at the boundary of the signal. For signals which have weak inherent laws and abnormal changes of boundary data, one can only consider the extreme points at the boundary of the signal. This makes the method more adaptive. Since power quality disturbances usually last for a while, power quality signals contain information of the disturbances at the endpoints of the signals. This feature satisfies the hypothesis of the adaptive waveform matching extension. So the improved HHT based on adaptive waveform matching extension is more suitable for power quality analysis.

The principle of the adaptive waveform matching method is shown in Fig. 3. In Fig. 3, $x(t)$ is the original signal; M_1 and N_1 are the first maximum and minimum values from the left point of the signal, corresponding to time tm1tm1 and tn1tn1, respectively; P_1 is the left boundary point. The P1–M1–N1P1–M1–N1 triangular waveform is used as the characteristic waveform. All sequences are searched for the most matching waveform Pi–Mi–NiPi–Mi–Ni. Data before (after) the matching waveform are taken as the left (right) extension of $x(t)$, so as to satisfy the natural trend of the signal. Conventionally, only the amplitudes of point P, point $M$and point N are considered in the waveform matching error computational formula. But for the description of a signal, in addition to the depth of the waveform above, rise time and fall time should also be considered. Therefore this paper improves the waveform matching error computational formula to takes all these factors into account to enhance the accuracy of the found matching waveform and to improve the end effect of HHT.

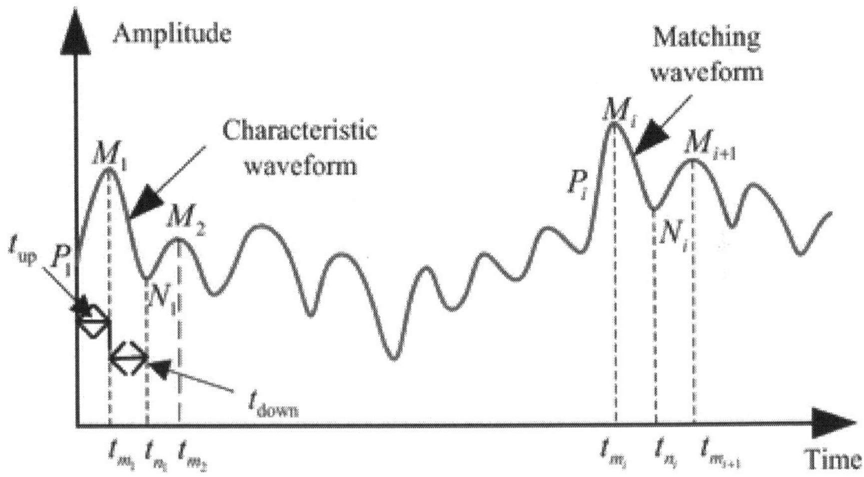

Figure 3. Principle diagram of adaptive waveform matching method Time

The steps of adaptive waveform matching method are shown as follows.

1) Calculate the corresponding time of boundary point P_i of the matching waveform by

$$t_{P_i} = \frac{t_{m_1} t_{n_i} - t_{n_1} t_{m_i}}{t_{m_1} - t_{n_1}} \tag{1}$$

2) Calculate matching error of characteristics waveform by

$$\begin{cases} \delta(i) = \delta_1(i) + \delta_2(i) \\ \delta_1(i) = |P_1 - P_i| + |M_1 - M_i| + |N_1 - N_i| \\ \delta_2(i) = \left|t_{\text{up},1} - t_{\text{up},i}\right| + \left|t_{\text{down},1} - t_{\text{down},i}\right| \end{cases} \tag{2}$$

where δ1(i)δ1(i) is the magnitude error of waveforms; δ2(i)δ2(i) is the error of the rise time and fall time of waveforms.

3) Get the new δ2(i)δ2(i) by the normalization processing of matching error, and then get the normalized matching error δ(i)δ(i) with new δ2(i)δ2(i).

$$\begin{aligned} \delta_2(i) &= \left|t_{\text{up},1} - t_{\text{up},i}\right| + \left|t_{\text{down},1} - t_{\text{down},i}\right| \\ &= \alpha|(M_1 - P_1) - (M_i - P_i)| \\ &\quad + \beta|(N_1 - M_1) - (N_i - M_i)| \end{aligned} \tag{3}$$

$$\begin{aligned} \delta(i) &= |P_1 - P_i| + |M_1 - M_i| + |N_1 - N_i| \\ &= \alpha|(M_1 - P_1) - (M_i - P_i)| \\ &\quad + \beta|(N_1 - M_1) - (N_i - M_i)| \end{aligned} \tag{4}$$

where t_{up} is the rise time of the signal and tup=tM−tPtup=tM−tP; tdowntdown is the fall time of the signal and tdown=tN−tM; *α* and *β* are constants which are obtained by (5). *M, N, P* are the corresponding vectors.

$$\begin{cases} \alpha = \dfrac{|M - P|}{|t_M - t_P|} \\ \beta = \dfrac{|N - M|}{|t_N - t_M|} \end{cases} \tag{5}$$

4) Take the waveform which has the smallest matching error in all matching errors as matching waveform.

5) Fetch data from the matching waveform between the previous point of the nearest data point to P_i and the previous extreme point to obtain the extension waveform and extend the waveform to the left endpoint.

6) Extend the signal at the right endpoint in the same way which is shown in step 5, and get the complete extension of the signal as x′(t)x′(t).

7) Apply HHT analysis to the extended x′(t)x′(t), and get the instantaneous frequency and amplitude of the signal.

3.2 End effect suppression simulation

Test of the short-time harmonic signal in (6) is conducted to validate the adaptive waveform matching extension method in end effect suppression. HHT analysis results of the test are given as follows.

$$u(t) = \sum_{i=1}^{2} A_i \sin(2\pi f_i t) \tag{6}$$

where A1=3; f1=30; A2=1.5; f2=5; the sampling frequency is 256 Hz; the number of sampling points is 512. The cubic spline interpolation is used to fit the envelopes of the signal.

The results in Fig. 4a indicate that with the extension method, the envelopes conform better to the natural trend of the signal. And the comparison of Fig. 4b with Fig. 4c shows that distortion at the endpoints, i.e., the end effect, is perfectly suppressed with the extension.

4. EXPERIMENTAL VALIDATIONS

In order to evaluate the performance of the improved HHT with adaptive waveform matching extension in power quality disturbance detection of microgrid, different simulations are implemented and the results are discussed. Improved HHT with adaptive waveform matching extension and HHT combining artificial neural network and mirror extension are compared below.

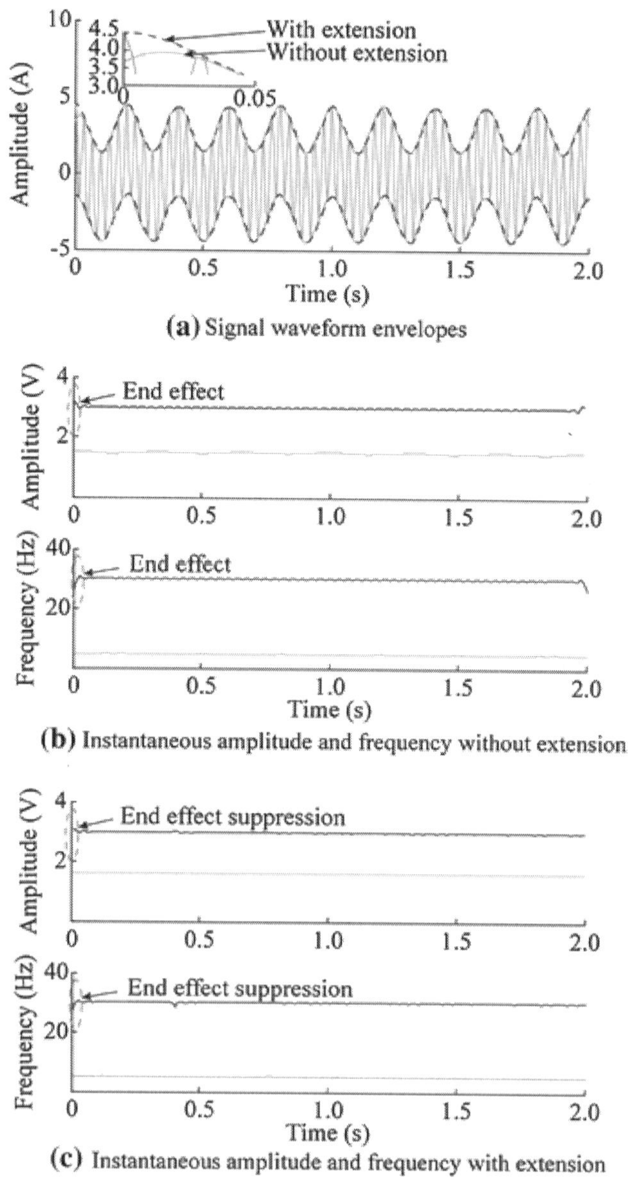

Figure 4. HHT analysis results of short-time harmonic signal

4.1 Voltage sag detection

In microgrid, sudden startup of wind turbines may cause voltage sags. The voltage sag signal given by (7) is extended directly since it is a single component signal.

$$u(t) = \begin{cases} \cos(2\pi ft) & \text{others} \\ 0.7\cos(2\pi ft) & 0.15 \leq t \leq 0.35 \end{cases} \tag{7}$$

As can be seen from Fig. 5, the voltage sag starts at t = 0.15 s and ends at t = 0.35 s, and the amplitude of the voltage disturbance is 0.3 V. Both two methods improve the end effect problem. The comparison between HHT combining artificial neural network and mirror extension is shown in Table 1. The improved HHT method can improve the amplitude accuracy by 0.52%, and the frequency accuracy is improved by 0.19%. We can also know that the detection time is reduced by 0.1037 s. That is to say, the detection precision and response speed of the method proposed in this paper is better. It is thus clear that the improved HHT method is accurate and valid for voltage sag detection in microgrid.

4.2 Integer harmonic detection

An integer harmonic signal is given as

$$u(t) = \sum_{i=1}^{3} A_i \sin(2\pi f_i t) \tag{8}$$

where A1=1, f1=50 of the fundamental wave; A2=0.3, f2=250 of the fifth harmonic component; A3=0.1, f3=350f of the seventh harmonic component. The sampling frequency is 3600 Hz, and the number of sampling points is 1800.

It can be clearly seen in Fig. 6 that the improved HHT extracts each harmonic component quiet well, and obtains precise instantaneous frequency and amplitude of each harmonic component. Both two methods improve the end effect problem. From Table 2, compared with the HHT combining artificial neural network and mirror extension, the improved HHT method proposed in this paper can improve the amplitude accuracy by 0.89% (5th) and 0.84% (7th), and the frequency accuracy is improved by 0.17% (5th) and 0.11% (7th). We can also know that the detection time is reduced by 0.0573 s. That is to say, the detection precision and response speed of the method proposed in this paper is better. It is thus clear that the improved HHT method proposed in this paper can accurately detect integer harmonics in microgrid.

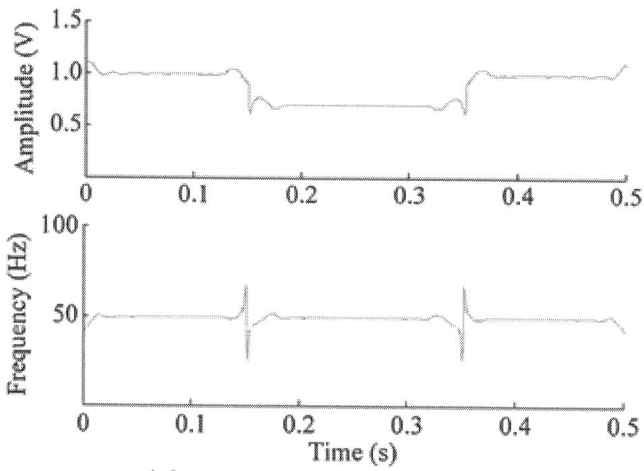

(a) Improved HHT combining artificial neural network and mirror extension

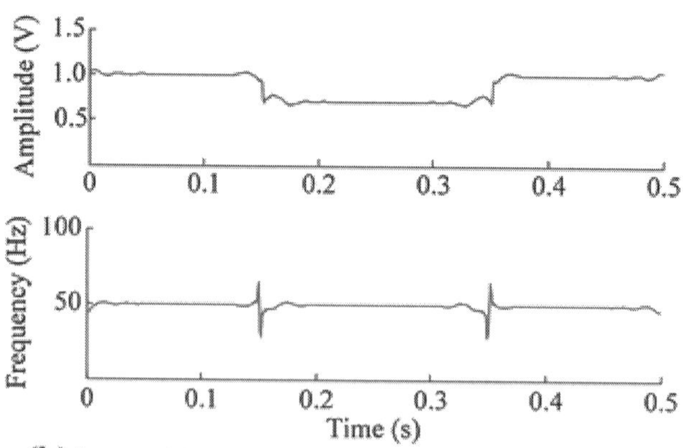

(b) Improved HHT with adaptive waveform matching extension

Figure 5. Improved HHT analysis results of voltage sag signal

Table 1. Results of two methods for voltage sag

Analysis method	Error rate of amplitude (%)	Error rate of frequency (%)	Time consuming (s)
Improved HHT combining artificial neural network and mirror extension	1.23	0.71	0.1037
Improved HHT with adaptive waveform matching extension	2.21	2.02	0.0450

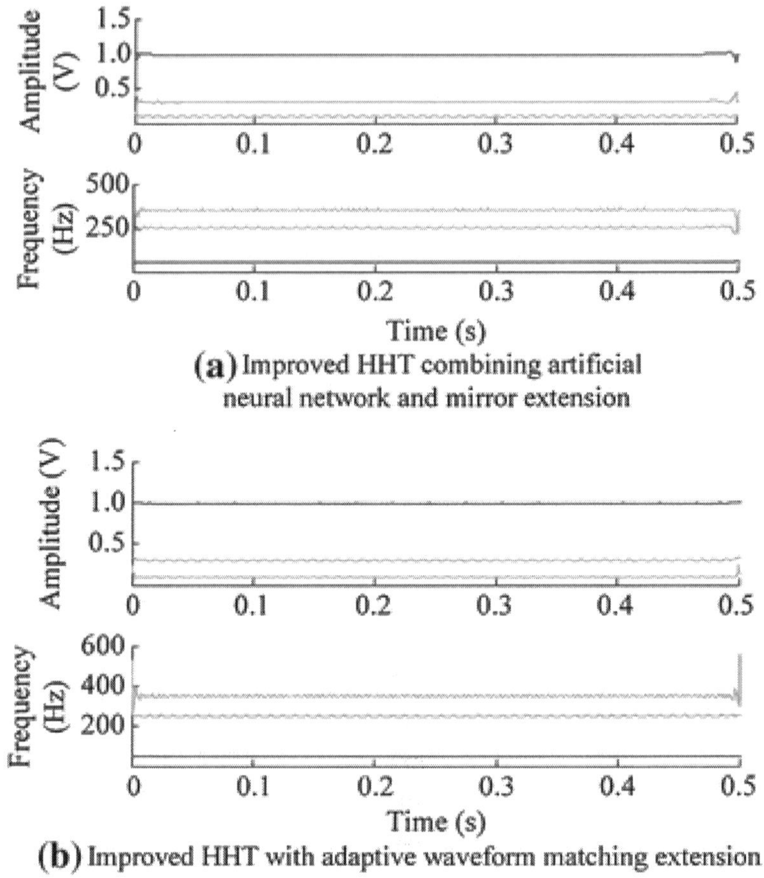

Figure 6. Improved HHT analysis results of integer harmonic signal

4.3 Inter-harmonic detection

Inter-harmonics which have the less or greater frequencies than the fundamental wave are selected for detection. The inter-harmonic signal is expressed as

$$u(t)=\sum_{i=1}^{3} A_i \sin(2\pi f_i t + \varphi_i) \tag{9}$$

Table 2. Results of two methods for integer harmonic

Analysis method	Error rate of amplitude (%)		Error rate of frequency (%)		Time consuming (s)
	5th	7th	5th	7th	
Improved HHT combining artificial neural network and mirror extension	1.22	0.95	0.21	0.14	0.0986
Improved HHT with adaptive waveform matching extension	0.33	0.11	0.04	0.03	0.0413

where $A_1 = 1,\ f_1 = 50,\ \varphi_1 = \pi/4$ for the fundamental wave, $A_2 = 0.5, f_2 = 25,\ \varphi_2 = \pi/15$ for the 0.5th harmonic component, $A_3 = 0.1,\ f_3 = 275,\ \varphi_3 = \pi/24$ for the 5.5th harmonic component. The sampling frequency is 2048 Hz, and the number of the sampling points is 1024.

From Fig. 7, we can know that both two methods improve the end effect problem. From Table 3, compared with the HHT combining artificial neural network and mirror extension, the improved HHT method proposed in this paper can improve the amplitude accuracy by 0.82% (5th) and 0.80% (7th), and the frequency accuracy is improved by 0.32% (5th) and 0.144% (7th). We can also know that the detection time is reduced by 0.0616 s. That is to say, the detection precision and response speed of the method proposed in this paper is better. It is thus clear that the improved HHT method proposed in this paper can precisely detect inter-harmonics in microgrid.

4.4 Voltage fluctuation and flicker detection

Since human's eyes are most sensitive to approximately 8 Hz voltage flickers, a simulation of the 8 Hz voltage fluctuation and flicker defined by (10) is carried out.

$$\begin{cases} u(t) = A(1 + Mp(t)\cos(2\pi f_1 t))\cos(2\pi f t) \\ P(t) = \begin{cases} 1 & 0.15 \leq t \leq 0.65 \\ 0 & \text{others} \end{cases} \end{cases} \tag{10}$$

where $A = 1,\ M = 0.1,\ f = 50\,\text{Hz},\ f_1 = 8\,\text{Hz}.$ The sampling frequency is 1024 Hz, and the number of sampling points is 1024. Accurate start time and end time of the voltage flick are detected as seen from Fig. 8.

From Fig. 8, we can know that both two methods improve the end effect problem. From Table 4, compared with the HHT combining artificial neural network and mirror extension, the improved HHT method proposed in this paper can improve the amplitude accuracy by 1.20%, and the frequency accuracy is improved by 0.51%. We can also know that the detection time is reduced by 0.0484 s. That is to say, the detection precision and response speed of the method proposed in this paper is better. Simulation results indicate that the improved HHT proposed in this paper is effective in the voltage flicks detection.

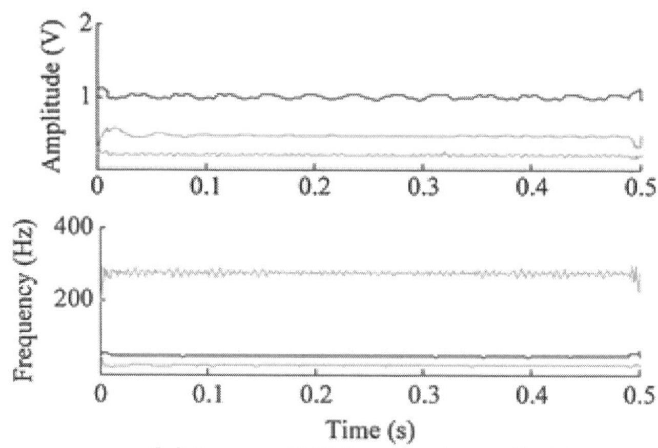

(a) Improved HHT combining artificial neural network and mirror extension

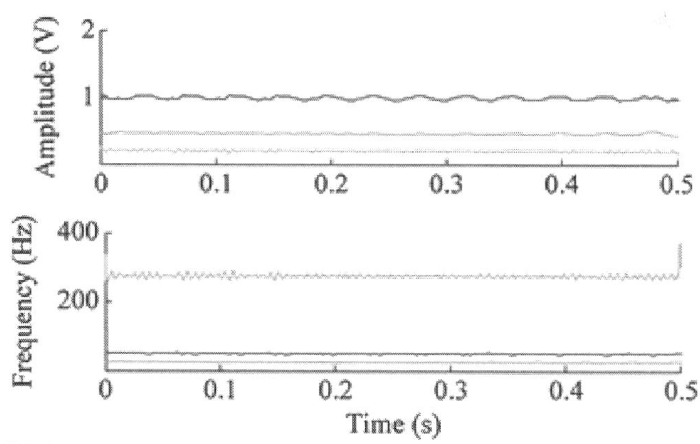

(b) Improved HHT with adaptive waveform matching extension

Figure 7. Improved HHT analysis results of inter-harmonic signal

Table 3. Results of two methods for inter-harmonic

Analysis method	Error rate of amplitude (%)		Error rate of frequency (%)		Time consuming (s)
	0.5^{th}	5.5^{th}	0.5^{th}	5.5^{th}	
Improved HHT combining artificial neural network and mirror extension	1.80	1.05	0.44	0.18	0.1024
Improved HHT with adaptive waveform matching extension	0.98	0.25	0.12	0.036	0.0408

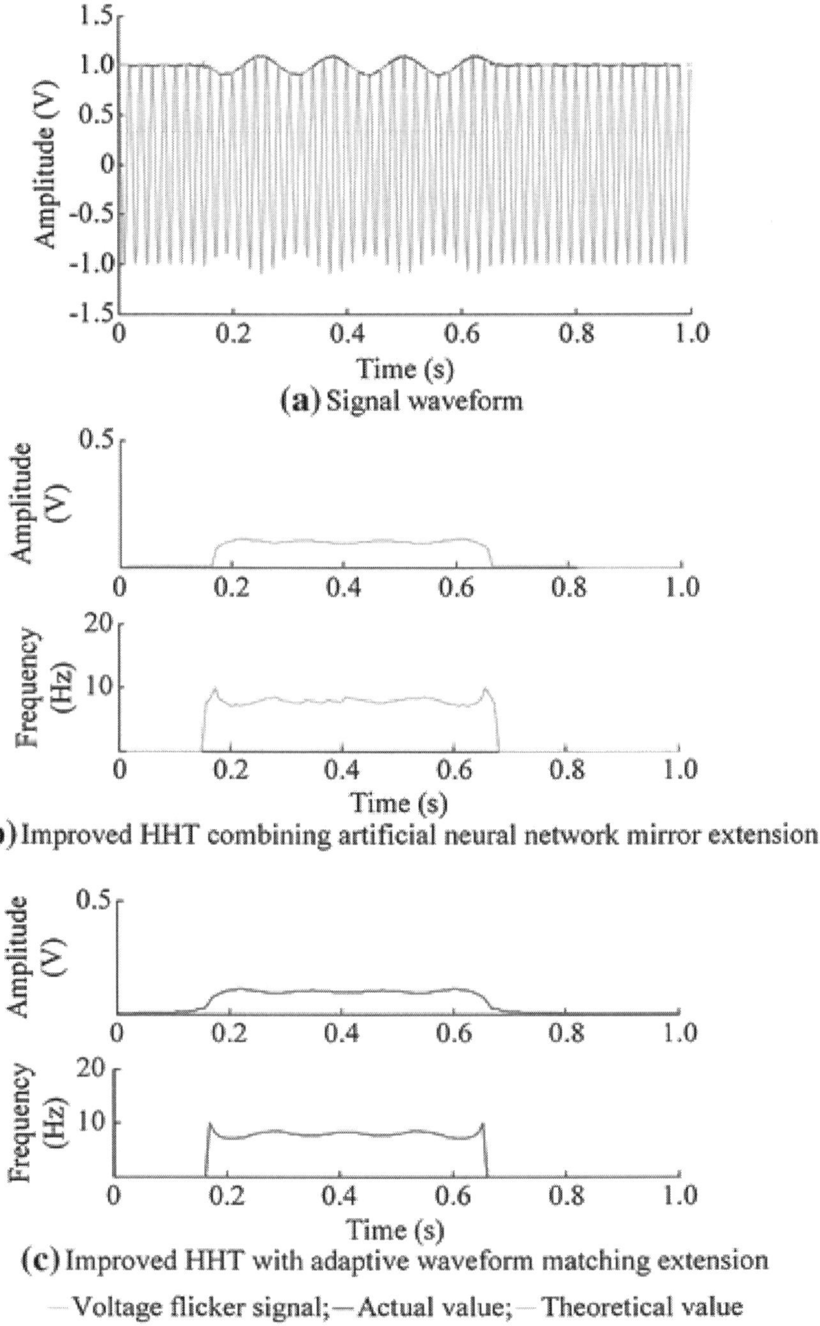

Figure 8. Improved HHT analysis results of voltage flicker signal

Table 4. Results of two methods for voltage flicker

Analysis method	Error rate of amplitude (%)	Error rate of frequency (%)	Time consuming (s)
Improved HHT combining artificial neural network and mirror extension	2.30	2.64	0.1016
Improved HHT with adaptive waveform matching extension	1.10	2.13	0.0532

5. ANALYSIS OF MEASURED RESONANCE WAVEFORM BASED ON IMPROVED HHT

5.1 Fluke harmonic experiment

A current waveform is generated by the Fluke harmonic generator as shown in Fig. 9. The sampling frequency is 12800 Hz, and 256 points are sampled in one sampling period.

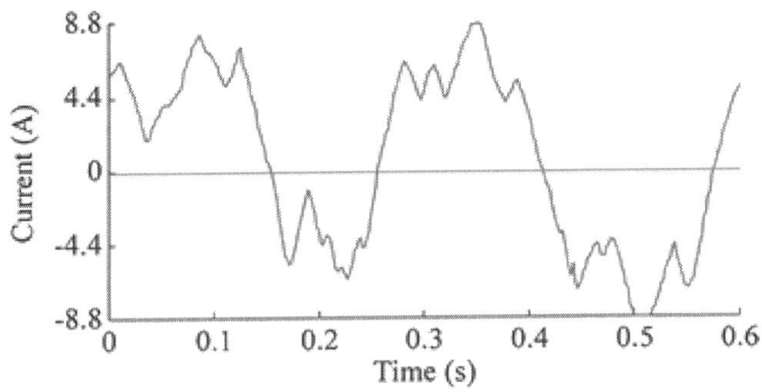

Figure 9. Current waveform of Fluke harmonic generator

Improved HHT analysis results in Fig. 10 show that the harmonic current of different frequency amplitudes including the 0.5^{th}, 3th, 5th, 5.5^{th}, 7th and 11th harmonic components, which is in accordance with the current waveform shown in Fig. 9. This indicates that the improved HHT works well for integer harmonics, fractional harmonics and inter-harmonics detection.

5.2 Field experiment of photovoltaic grid-connected system

When SVG reactive power compensation devices and all devices in the PV field area were at normal operation, the 110 kV bus voltage had been measured through a quality recorder (USA Fluke 1760) for 7 days (September 26th, 2013 - October 3rd, 2013) in a 30 MWp photovoltaic grid-connected system of an industrial park. The voltage distortion of phase A at the period of September 26th 15:00-21:00 which is obtained by power quality analyzer is taken as an example in Fig. 11. The improve HHT analysis results are shown in Fig. 12. The vertical axis represents the total voltage harmonic distortion (THD) percentage value.

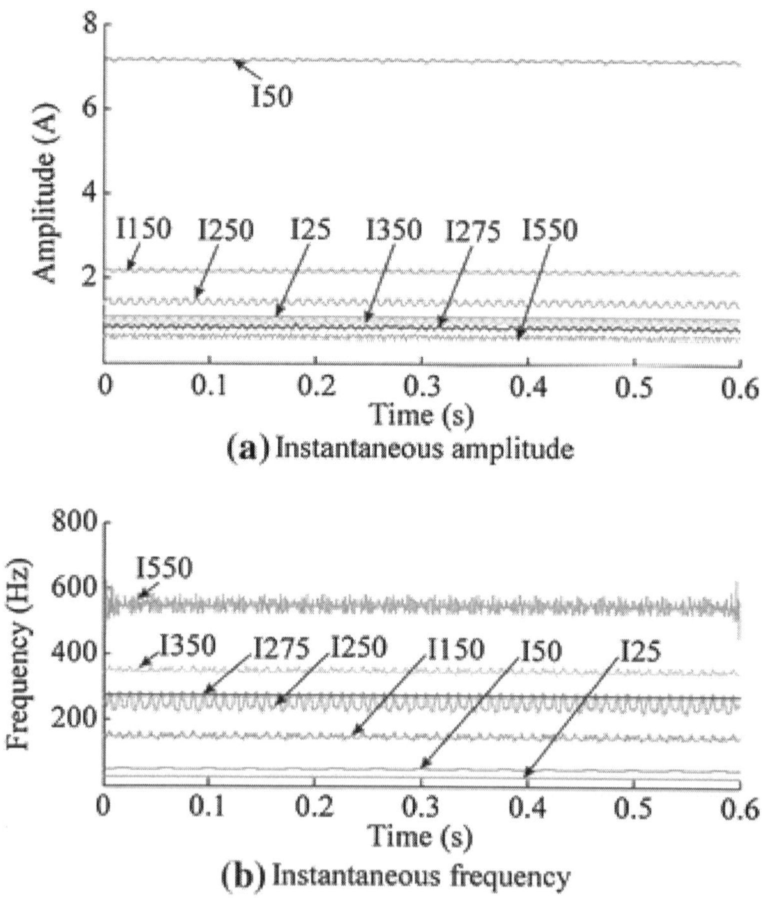

Figure 10. Improved HHT analysis results of harmonic currents

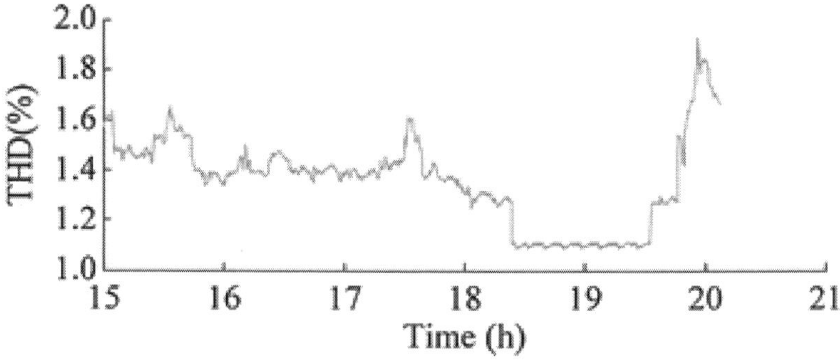

Figure 11. THD of field phase A voltage distortion

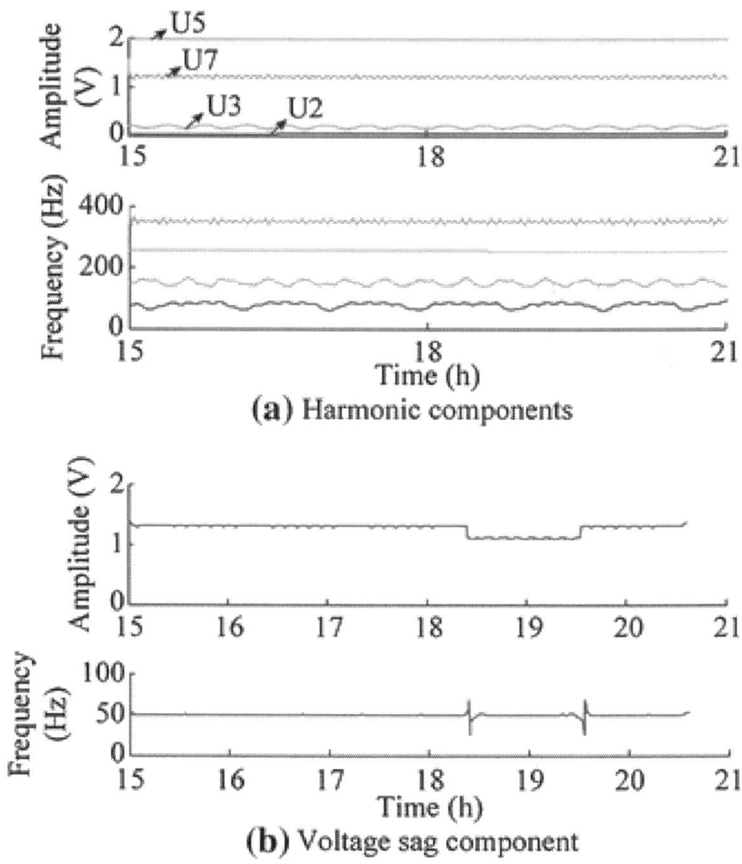

Figure 12. Improved HHT analysis results of field voltage distortion

As can be seen from Fig. 12, the voltage harmonic distortion contains the 2th, 3th, 5th, 7thharmonic and voltage sag components. The harmonic orders, harmonic contents and voltage sag information obtained by the proposed method in this paper are consistent with the results of the field measurement obtained by power quality analyzer. Hence, the effectiveness of the proposed method is verified.

6. CONCLUSIONS

This paper systematically analyzes the causes of the end effect of HHT and the mode mixing problem. An improved HHT with adaptive waveform matching extension is proposed. This innovative method considers not only the depth of the waveform, but also the rise time and fall time of the waveform. Thus it is more accurate than traditional waveform extension method. Both simulation and experiment results show that adaptive waveform matching extension can effectively suppress the end effect of original HHT and enhance the accuracy of HHT analysis. The improved HHT is valid in various power quality disturbance detection of microgrid. Moreover, it provides a theoretical basis for the wider application of HHT in power quality analysis.

As perspective, the proposed improved HHT method may be used to detect sudden variation of different power quality index of a microgrid and to give right information to energy management system in order to determine appropriate working modes (grid-connected or islanded).

ACKNOWLEDGMENTS

This work was supported by National High Technology Research and Development Program of China (863 Program) (No. 2015AA050104) and National Natural Science Foundation of China (No. 51577068).

REFERENCES

1. Shi T, Cui HF, Huang RM (2012) Applied research of battery energy storage in grid-connected photovoltaic power. In: Proceedings of the 2012 world automation congress (WAC'12), Puerto Vallarta, 4 pp. Accessed 24–28 June 2012

2. Tayjasanant T, Wang WC, Li C et al (2005) Interharmonic-flicker curves. IEEE Trans Power Deliv 20(2):1017–1024

3. Chang GW, Chen CI (2010) An accurate time-domain procedure for harmonics and interharmonics detection. IEEE Trans Power Deliv 25(3):1787–1795

4. Chen CI, Chang GW (2010) Virtual instrumentation and educational platform for time-varying harmonic and interharmonic detection. IEEE Trans Ind Electron 57(10):3334–3342

5. Antonino-Daviu JA, Riera-Guasp M, Pineda-Sanchez M et al (2009) A critical comparison between DWT and Hilbert–Huang-based methods for the diagnosis of rotor bar failures in induction machines. IEEE Trans Ind Appl 45(5):1794–1803

6. Huang NE, Shen Z, Long SR (1999) A new view of nonlinear water waves: the Hilbert spectrum. Ann Rev Fluid Mech 31(1):417–457CrossRefMathSciNet

7. Huang N, Shen Z, Long SR et al (1998) The empirical mode decomposition and the Hilbert spectrum for nonlinear and non-stationary time series analysis. Proc R Soc A 454:903–995

8. Zhang WB, Wang C, Wang HX (2012) Hilbert-Huang transform-based electrostatic signal analysis of ring-shape electrodes with different widths. IEEE Trans Instrum Meas 61(5):1209–1217

9. Wu ZH, Huang NE (2010) On the filtering properties of the empirical mode decomposition. Adv Adapt Data Anal 2(4):397–414

10. Lin SC, Li PC (2011) Automatic contrast enhancement using ensemble empirical mode decomposition. IEEE Trans Ultrason Ferroelectr Freq Control 58(12):2680–2688

11. Cheng YC, Li ST (2012) Fuzzy time series forecasting with a probabilistic smoothing hidden Markov model. IEEE Trans Fuzzy Syst 20(2):291–304

12. Wei L, Guo W, Wen F et al (2010) Waveform matching approach for fault diagnosis of a high-voltage transmission line employing harmony search algorithm. IET Gener Transm Distrib 4(7):801–809

13. Wang XZ, Wang JP (2012) Research of fault diagnosis based on matching pursuit and biomimetic pattern recognition. In: Proceedings of the 10th world congress on intelligent control and automation (WCICA'12), Beijing, pp 4848–4852. Accessed 6–8 July 2012

14. Yang YL, Deng JH, Wu CP (2009) Analysis of mode mixing phenomenon in the empirical mode decomposition method. In: Proceedings of the 2nd international symposium on information science and engineering (ISISE'09), Shanghai, pp 553–556. Accessed 26–28 Dec 2009

15. Hu XY, Peng SL, Hwang WL (2012) EMD revisited: a new understanding of the envelope and resolving the mode-mixing problem in AM-FM signals. IEEE Trans Signal Process 60(3):1075–1086

16. Wu ZH, Huang NE (2009) Ensemble empirical mode decomposition: a noise-assisted data analysis method. Adv Adapt Data Anal 1(1):1–41

17. Deering R, Kaiser JF (2005) The use of a masking signal to improve empirical mode decomposition. In: Proceedings of the IEEE international conference on acoustics, speech, and signal processing (ICASSP'05), vol 4, Philadelphia, pp 485–488. Accessed 18–23 Mar 2005

Chapter 5

HOS network-based classification of power quality events via regression algorithms

José Carlos Palomares Salas[1,2*], Juan José González de la Rosa[1,2], José María Sierra Fernández[1,2] and Agustín Agüera Pérez[1,2]

[1]Research Group PAIDI-TIC-168: Computational Instrumentation and Industrial Electronics (ICEI), Av. Ramón Puyol S/N., E-11202 Algeciras-Cádiz, Spain

[2]Area of Electronics, Polytechnic School of Engineering, University of Cádiz, Av. Ramón Puyol S/N., E-11202 Algeciras-Cádiz, Spain

ABSTRACT

This work compares seven regression algorithms implemented in artificial neural networks (ANNs) supported by 14 power-quality features, which are based in higher-order statistics. Combining time and frequency domain estimators to deal with non-stationary measurement sequences, the final goal of the system is the implementation in the future smart grid to guarantee compatibility between all equipment connected. The principal results are based in spectral kurtosis measurements, which easily adapt to the impulsive nature of the power quality events. These results verify that the proposed technique is capable of offering interesting results for power quality (PQ) disturbance classification. The best results are obtained using radial basis networks, generalized regression, and multilayer perceptron, mainly due to the non-linear nature of data.

KEYWORDS

Artificial neural networks (ANN), Power quality (PQ), Cumulants, Higher-order statistics (HOS), Regression algorithms, Smart grid (SG), Spectral kurtosis (SK)

1. INTRODUCTION

With the consequent unstoppable increase of electronic equipments demanding electricity, consumers expect uninterrupted availability and quasi perfect power quality (PQ). For that reason, PQ is being the object of continuous interest by researchers and developers, due to its influence over and from the loads, and the recent potential inclusion in the modern smart grid (SG).

In this frame, an adequate PQ assures the necessary compatibility between all equipment connected to the grid [1]. The terms of this compatibility gather several aspects: sustainable power with low losses and high quality and security of supply and safety, being at the same time economically efficient, reliable, and resilience [2,3].

Certainly, the future SG would introduce transformative technologies to meet these design requirements, integrating intelligence into end-use devices as the key to satisfying the demand response. In parallel, the industry would design ways to incorporate automatic end-use-load participation into the model so that customers are not bothered by these programs and decisions, and so their lifestyles are not inconvenienced [4].

Due to the above arguments, the role of smart meters and sensors is the first being revised in the present and future SG. These automated meters (AM) used a two-way communicating infrastructure and centralized management, as well as new features such as the following: outage management, demand response, automatic load shedding, distribution automation, and the ability to enable and commute alternative energy sources.

Provided with this scenario, this research integrates artificial neural networks (ANNs) and advanced signal processing techniques based in higher-order statistics (HOS) in order to be implemented into an automated smart meter for PQ event detection and classification, within the frame of a SG with a high distribution penetration of renewable sources. Seven regression algorithms are tested based on an hybrid time-frequency battery of characteristics, specially designed to deal with non-stationary measurement time series.

The feature extraction stage from PQ disturbances is based on HOS, idea which has been proven to be efficient in several works. Indeed, since PQ events are sudden changes in the power line, the HOS are potentially useful to characterize each type of electrical anomaly both in time and frequency domains. As a novelty, the present research exploits the combination of time and frequency domain features, to deal with the inherent non-stationary associated to the electrical anomalies, with the goal to improve the performance of the ANN and to make feasible the integration in an smart meter.

Regarding backgrounds of HOS applications in this field, in the time domain, several notable works are worthy, e.g., Bollen et al. introduced new advanced statistical features to PQ event detection [5]. In the same direction, Gu and Bollen [6] found relevant characteristics associated to PQ events in the time and frequency domains. The work by Ribeiro et al. is also remarkable [7], which extracted new time-domain features based in cumulants. The same authors performed the classification of single and multiple disturbances using HOS in the time domain and Bayes' theory-based techniques [8]. HOS techniques and estimators have also been implemented to specifically detect sags and swells [9].

The categorization of PQ anomalies had been formerly performed by Nezih and Ece in the work [10], where they proved that HOS and quadratic classifiers improve the second-order-based methods. The same authors previously achieved performance in second-order computing, using 2-D wavelets and compression techniques [11,12]; finding, despite the promising results, the limits of the procedure and quantifying its heavy computational cost. Alienated to this work, the researches of Poisson et al. and Santoso et al. [13,14] also reported a wavelet-based method, finding the potential and the drawbacks of the technique so that to implement it in an intelligent meter.

The direct antecedent of the present research in the work by J.J.G. de la Rosa et al. [15], in which they performed a mixed study involving the time-domain variance, skewness, and kurtosis, and they obtained consequences combined with the spectral kurtosis (SK), and over a set of real-life measurements, some of them with mixed PQ perturbations. The same authors proposed a preliminary criteria for seven types of disturbances based on the former estimators [16]. In a previous work [17], they designed an offline case-based reasoner based on time-domain HOS estimators. Furthermore, the authors used also HOS features in classification techniques for characterization of electrical PQ signals [18].

The present paper is designed as follows. The next section summarizes the main advances in the field, paying special attention at the applications of ANN for PQ analysis, and reasoning the contribution of HOS to the feature extraction stage and the architecture. Then, Section Higher-order statistics for PQ monitoring: an enhancement proposal over the DWT exposes the advantages of HOS for PQ monitoring and the state-of-the art. In SectionProposed methodology: the HOS-based

ANN, the procedure is detailed in order to expose the results later in Section Results; finally, conclusions are drawn in Section Conclusions.

2. ANN FOR PQ ANALYSIS: TOWARDS THE HOS PARADIGM

The present paper postulates non-stationary signal processing with higher-order statistics in the time and frequency domains in order to extract a battery of features to be processed via ANNs with regression algorithms. ANNs have been used for classification purposes in myriads of works and have proved the utility for a long time [19]; Figure 1shows a generic architecture, indicating how the input pattern is processed via the neurons to provide with the output vector, in order to introduce the concrete architecture in the present work.

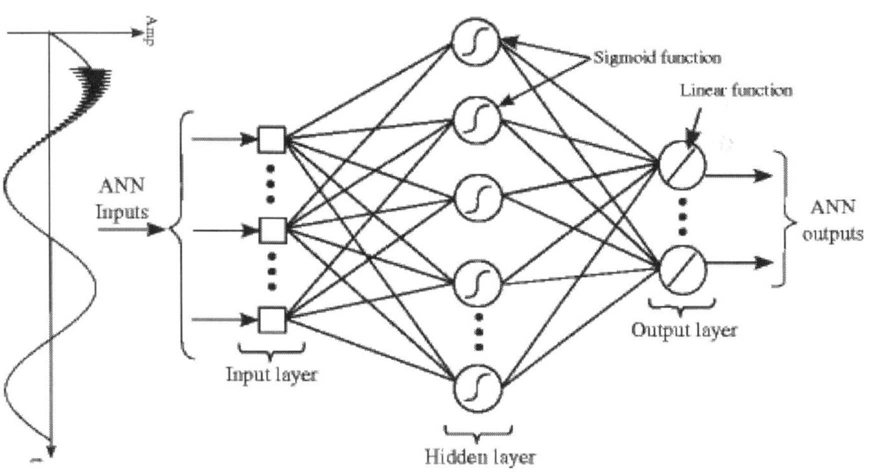

Figure 1. Generic ANN structure which postulates the feature extraction stage previous to the ANN.

The performance of the potential PQ monitoring system and consequently the ANN is directly related to the pre-processing and feature extraction techniques used. The main goal of the feature extraction is to represent the data set in a new feature space in which the probability to distinguish classes is higher than the one in the original space. Therefore, the identification of efficient pre-processing and feature extraction techniques is a key issue [20].

Some works have shown that the best performance for non-stationary signals is achieved combining time and frequency characteristics obtained in the feature extraction stage via wavelet transforms (WT), on the hypotheses that WT performance is suitable for the analysis of transient signals with easy computation. Three works are worth to be cited as they are directly concerned with our research, involving ANN and PQ analysis. In the first one to mention, Angrisani et al. presented a wavelet network for automated transient detection; basically, it was based in an extended perceptron which incorporated wavelet nodes [21]. They proposed a unique structure based on the introduction of wavelet nodes in the traditional ANN, in the neurons of the first layer, and was applied successfully to typical transients in the PQ analysis. The number of nodes was fixed to 12, and this number was related to the time location of the perturbation. The network has to be pre-conditioned in order to detect concrete transient signals.

In the same former line, using the WT as a preprocessing tool for ANN, the second work by Iñigo Monedero et al. [22] developed a neural perceptron-based (three hidden layers) real-time first system prototype, which was trained using a disturbance generator for pseudo-synthetic waveforms. They obtained a performance of 89% success combining simulated and real-time data. The perturbations were classified in frequency, voltage, and harmonics, and the topology of the ANN (number of hidden neurons and number of outputs) was highly moveable depending on the PQ event under study.

The third and most recent paper, by Martin Valtierra-Rodriguez et al. [23] involves a new dual neural-network-based methodology to detect and classify single and combined PQ disturbances, consisting, on the one hand, of an adaptive linear network for harmonic and inter-harmonic estimation that allowed computing the root-mean-square voltage and THD indices, from which it is possible to detect and classify sags, swells, outages, and harmonics. A complementary feed-forward neural network was used for pattern recognition using the horizontal and vertical histograms of a specific voltage waveform to classify spikes, notching, flicker, and oscillatory transients. The complementary action of these neural networks allows the detection and classification even with simultaneous electrical anomalies, both in noisy and noise less scenarios.

Apart from the difficulties in implementing a real-time processor in the smart grid frame, the main disadvantage of the above procedures lies in the fact that accuracy and repeatability are highly compromised by the second-order estimators (WT and V_{RMS}) used in the feature extraction stage. Drawbacks arise when the data are corrupted by noise; specially when the number of samples of the signal window is reduced, and the resolution and repeatability are degraded. This facts are dramatically increased if the tests are performed over synthetic signals of controlled-lab experiences, where predictability is tacitely supposed. In the real-world experiences, the system should be prepared for unpredictable phenomena, both in the time and in the frequency domains. As explained hereinafter, this goal is accomplished by statistical parameters of an order higher than two.

3. HIGHER-ORDER STATISTICS FOR PQ MONITORING: AN ENHANCEMENT PROPOSAL OVER THE DWT

During the last decade, some researches [7,15,24] have demonstrated the usage of HOS features for PQ monitoring. The motivation of HOS in PQ analysis is twofold. By one side, as HOS measurements are correlations that involve powers higher than two [25], the HOS information for Gaussian signals is null, as Gaussian processes are described up to the second order. As many measurement noises are Gaussian or symmetrically distributed, the HOS may be less degraded by the background noise than the second-order calculations. Secondly, the capability to reveal non-linear characteristics from the data, which is important for pattern recognition scenarios, in concrete to target PQ events, waveforms which exhibit high time variability, like impulsiveness of peakedness.

The usage of HOS as a feature extraction technique for PQ monitoring systems is very promising, and several recent works presented good results with respect to both detection and classification tasks. The most similar work was recently committed by Liu et al. [26]; transients were classified according to the features extracted via the SK and using ANNs. They worked with simulated signals over the five types of synthetic perturbations.

Formerly, it had been shown that combining techniques allows efficient classification of single and simultaneous disturbances, and more, the usage of the second- and fourth-order HOS features, for a specific lag chosen from Fisher's discriminant ratio (FDR) criterion, has been enough to deal with the majority of the disturbances considered [24].

Regarding specifically signal processing issues, some results show that HOS are capable of detecting disturbances even using short acquisition time windows, which represents an important characteristic for several power system applications such as protection, signal segmentation, and disturbance localization. Being specific, the results shown that the detection of disturbances can be accomplished in less than a quarter of cycle, which is excellent for protection application, where speed and accuracy need to be combined to guarantee selectivity and reliability during the occurrence, for example, of a fault in a system [24,25].

Regarding the HOS that has been used in the present research, it is worth remarking that we have made use of a combination in the time and the frequency domains. The ensemble of Equation 11, whose rigorous mathematical treatment is in the appendix, constitutes indirect measurements of the variance, skewness, and kurtosis that has been used in the present paper, along with the SK, which is used to locate the transients in the frequency domain.

With the aim of motivating the performance of HOS over second-order methods, a comparison has been developed via the discrete wavelet transform (DWT), which is a promising technique for PQ analysis according to the literature (e.g., [21]). Without the further necessity of expanding expressions, in this section, we recall that every finite energy signal $s(t)$ can be decomposed over a wavelet orthogonal basis according to:

$$s(t) = \sum_{j=-\infty}^{+\infty} \sum_{k=-\infty}^{+\infty} \langle s, \psi_{j,k} \rangle \psi_{j,k}. \quad (1)$$

Each partial sum, indexed by k, in Equation 1 represents the detail variations at the scale $a=2^{j}$ (at each level j, the scale is increased by a factor of two):

$$d_j(t) = \sum_{k=-\infty}^{+\infty} \langle s, \psi_{j,k} \rangle \psi_{j,k} \qquad s(t) = \sum_{j=-\infty}^{+\infty} d_j(t). \quad (2)$$

The approximation of the signal $s(t)$ can be progressively improved by obtaining more levels, with the aim of recovering the signal selectively. For example, if $s(t)$ varies smoothly, we can obtain an approximation by removing fine scale details, which gather information regarding the high frequencies or rapid variations of the signal. This is done by truncating the sum in Equation 1 at the scale $a=2^{J}$:

$$s_J(t) = \sum_{j=J}^{+\infty} d_j(t). \quad (3)$$

Details corresponding to indexes $j < J$ are not considered in Equation 3.

Hereinafter, a simple experience is exposed illustrating the limitations of the method based in the DWT, in case DWT coefficients were selected to be used as features. The wavelet decomposition tree has been expanded up to level 7 using the mother wavelet *sym8*. The idea in the background literature is to study the distribution of the signal energy for each level of the tree, which acts as a filter bank. As stated and proved in myriads of references (e.g., [27]), low index levels point rapid fluctuations of the signal, whereas slow variations are associated to higher (or deeper) decomposition branches. The difference between the energy of the perfect power signal and the energy of the PQ event under test has been studied for each level. The energy expression that we have considered is the traditional 2-norm, according to Equation 4:

$$E(\mathbf{s}) := \sum_{i=1}^{N} \|s_i\|^2, \tag{4}$$

where $\mathbf{s}$ is the N-point vector signal and s_i, their associated components. Figure 2 shows the processing results for six prototypes. The limitations of the DWT can be concluded via a simple analysis of the graphs. Firstly, as the DWT analysis consists of a selective filtering, sag and swells cannot be distinguished. Secondly, mixed PQ events are also difficult to separate because the constant incremental energy does not indicate whether a transient is coupled or not. Finally, an eventual noisy environment would mask results in the case of the oscillatory transient (this has been extensively proved).

At the light of the second-order analysis, a more robust set of features is required in order to process real-life measurements. This is achieved via HOS.

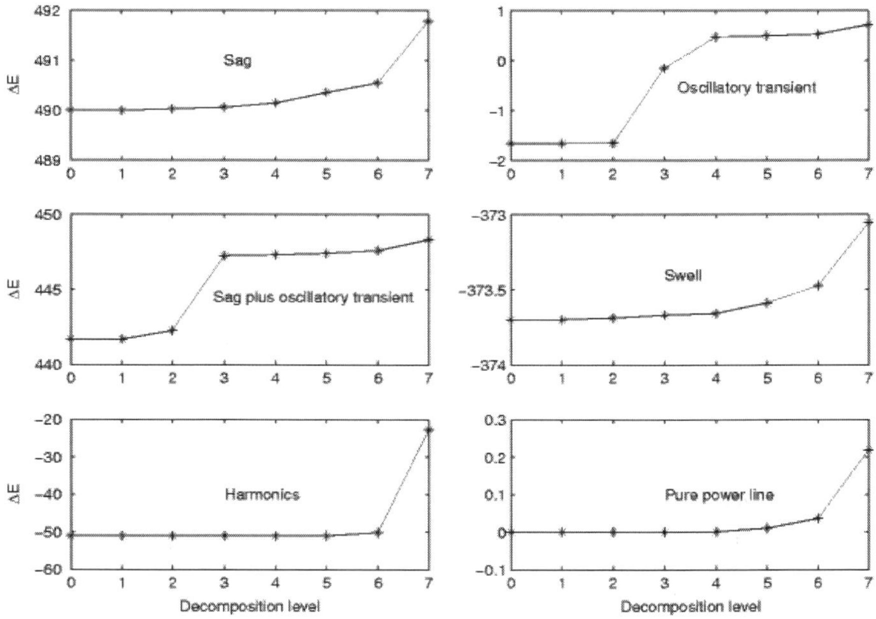

Figure 2. DWT analysis for different PQ events. From top to bottom and left to right: sag, oscillatory transient, sag plus oscillatory transient, swell, harmonics, and pure power signal. The energy difference is depicted vs. the decomposition level. In the case of additive noise, the method does not guarantee success in the feature extraction stage.

4. PROPOSED METHODOLOGY: THE HOS-BASED ANN

The final goal of this work is the implementation in the future SG to guarantee compatibility between all equipment connected. Data used in this work have been simulated using MATLAB software. The data set generated by the simulation consists of 550 samples including the different studied disturbance kinds, which cover the following disturbances: oscillatory transient, impulsive transient, interruption, harmonic permanent distortion, harmonic temporal distortion, sag, sag plus oscillatory transient, and swell. Each signal comprises a 20K-point synthetic time-domain register with a duration of 1 s (20 KHz sampling frequency). An additive normal noise process (1% of the amplitude of the signal) has been added in order to achieve a more realistic behavior. Figure 3 shows an example of these signals as well as healthy signal that are utilized in this paper.

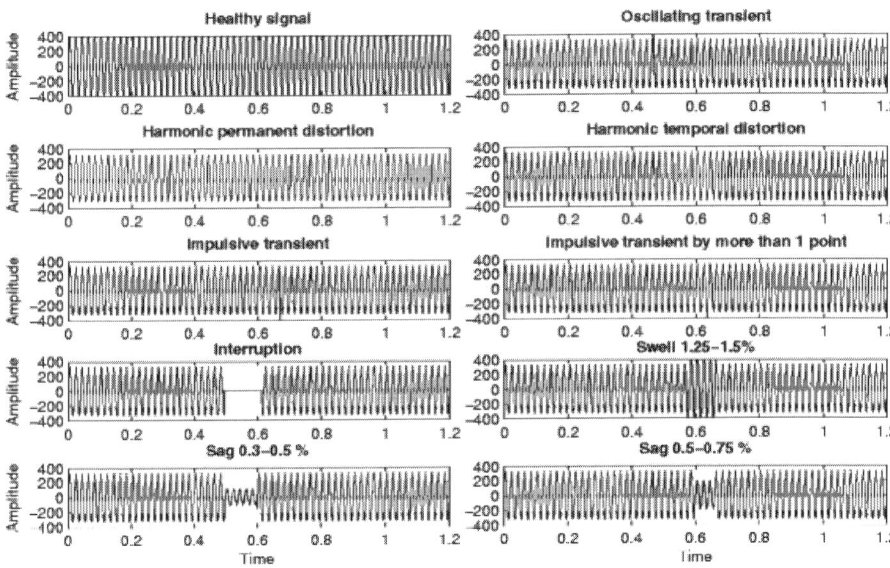

Figure 3. Example of healthy signal and different disturbances studied.

The classification techniques used to classify the PQ disturbances are based on regression algorithms. These techniques are multiple linear regression (MLR), adaptive linear neuron (LIN), multilayer ANNs (BP1 and BP2), radial basis function (RBF) network, exact radial basis (ERB), and generalized regression network (GRN). Every model has been intelligently adapted to meet the objective of PQ classification because each has different characteristics. Table 1 shows the selected parameters corresponding to the architecture and activation function of each model used. The final design of each model has been obtained by the optimization of the parameters shown in this table.

Table 1. Parameters of the ANNs

ANN	LIN	BP1	BP2	RBF	ERB	GRN
Hidden layers	-	1	2	1	1	1
Neurons hidden layer 1	-	[4 to 10]	[4 to 10]	[1 to 150]	[1 to 150]	[1 to 150]
Neurons hidden layer 2	-	-	[2 to 5]	-	-	-
TF	-	S	S	G	G	G
TF output	L	L	L	L	L	L
Training algorithm	WH	LM	LM	*k*	*k*	*k*
Spread	-	-	-	[1 to 20]	[1 to 20]	[1 to 20]

LM, Levenberg Marquardt; WH, Widrow Hoff; S, sigmoid; G, Gaussian; L, linear; *k*, *k*-means; TF, transfer function.

Data used to realize the classification are based on representative coefficients obtained from the PQ disturbances referred above. These coefficients are acquired by a process of feature extraction which is based on the combination of higher-order statistics in time and frequency domains. The HOS have been computed using a 400-points sliding window (which corresponds to a signal period), with a shift of 10 points over a vector of 20,000 points. After extraction stage, a total of 14 characteristic features are selected, nine of whom correspond to time domain and the remaining five to frequency domain. The coefficients selected in the first one correspond to the maxima and minima and stable in the second-, third-, and fourth-order cumulants at zero lags (directly related to the variance, skewness, and kurtosis). On the other hand, the coefficients selected in the second one correspond to the frequency of extreme value of SK, bandwidth of dome, extreme value of SK, number of peaks in SK, and dome very targeted (between 0 and 1). The abstract graphic of this work with the proposed methodology can be seen in Figure4 where it specifies the feature extraction module. In order to illustrate the capability of the SK to discriminate PQ disturbances, we have selected a practical example consisting of an oscillatory transient coupled to the power sine wave. The analysis result is depicted in Figure 5. The time-domain variance increases when it bumps into the transient; this behavior is independent of the transient frequency. Similarly, the time-domain skewness and kurtosis detect slight variations. The real detection takes place in the frequency domain; the SK produces a real enhancement in 2,000 Hz, along with the high-resolution bump (narrow peak).

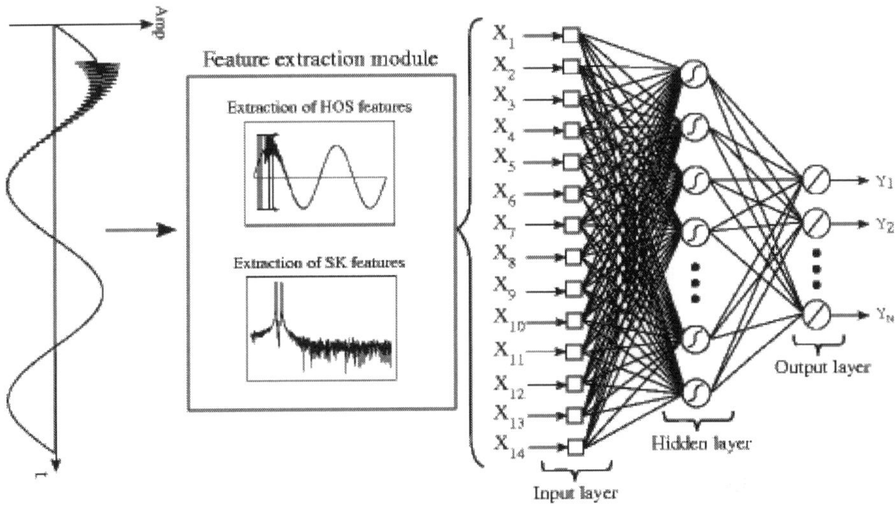

Figure 4. Procedure of feature extraction.

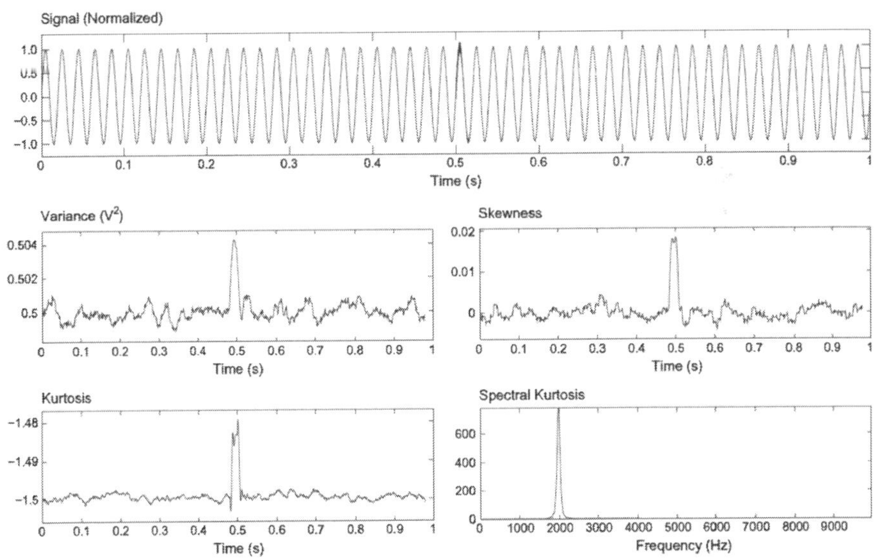

Figure 5. An example of mixed analysis in the time and frequency domains. An oscillatory transient is detected in the time domain by the second-, third-, and fourth-order statistics, and its frequency is clearly targeted by the spectral kurtosis.

The resulting data after carrying out the feature extraction is a matrix of dimension 550×14 (samples × features). Then, the building of the models is performed by following two steps in order to efficiently classify the disturbances. First, data are normalized so that they are in the interval [−1,1], for a faster computation [28]. And second one, we divide randomly the data set into three subsets: training, evaluation, and test sets. The training and validation sets, with 70% and 15% of data, respectively, were used for ANN model building; and the third set, with the last 15%, was used to test on the out-of-sample set the classification power of a model.

For comparison purposes, the classification criterion is based on the parameter hit rate (HR), that is defined as follows:

{HR}=NCNT×100%{HR}=NCNT×100% (5)

where N_T is the number of test samples and N_C is the number of correct disturbance recognition.

5. RESULTS

Once the assessed models were configured and optimized, they are used in the out-of-sample set. Because the database is small, we choose randomly two test sets, and in each of them, 100 experiments are launched by model. This is realized to achieve statistically meaningful results which rule out the random factors influencing the ANNs.

For each model, we have performed three analysis in function of the features used: HOS in the time domain (HOS_t), the SK, and the mixed (time and frequency domains) analysis (HOS_t+SK). In all of them, the HR was calculated for each model on the 200 offline tests, observing the arithmetic mean. The obtained results showing the percentage of effectiveness for classifying disturbances are presented in Table 2. In most models, results obtained in the mixed analysis are better than in the other analyses except isolated results. These exceptions represent the 27.27% of the cases presented in Table 2, where the 14.28% of them correspond to (HOS_t) features, and the remaining 12.99% correspond to SK features.

As can be seen, the algorithms based on linear models are worse than those based on non-linear schemes. The best models both individually and collectively are ERB, GRN, and BP1.

6. CONCLUSIONS

In this paper, seven regression algorithms have been applied and compared for PQ disturbances classification. The novel aspect is the introduction of new representative coefficients based on HOS in time and frequency domains. These coefficients are the inputs used in the classification algorithms to verify the occurrence or not of single or multiple disturbances in the electric signals.

The data used to test the proposed method were generated by the MATLAB software. The PQ disturbances considered are the most common on the supply. The best models both individually and collectively are obtained employing radial basis networks, generalized regression, and multilayer perceptron. The overall hit rates obtained are 94.70%, 79.59%, and 74.17%, respectively. This is consistent with non-linearity of the used data and emphases of the non-linearity that provide the HOS.

Once the obtained results demonstrate that the proposed method can effectively classify different kinds of PQ disturbances, is necessary to do more training and tests to find an ANN classifier with better characteristics than obtained in this research.

Table 2. Regression algorithms for ANNs

	MLR			LIN			BP1			BP2			RBF			ERB			GRN		
	HOS_t + SK	HOS_t	SK	HOS_t + SK	HOS_t	SK	HOS_t + SK	HOS_t	SK	HOS_t + SK	HOS_t	SK	HOS_t + SK	HOS_t	SK	HOS_t + SK	HOS_t	SK	HOS_t + SK	HOS_t	SK
Harmonic permanent distortion	42.45	15.22	0.00	0.00	0.00	0.00	85.89	59.88	88.00	78.30	49.61	70.00	54.94	52.87	100.00	100.00	83.22	100.00	69.39	1.83	0.00
Harmonic temporal distortion	34.14	29.44	4.70	10.54	0.00	1.30	74.56	60.73	17.12	70.00	59.91	24.82	42.17	53.02	1.79	93.00	79.71	80.58	48.41	33.77	0.00
Impulsive transient	65.71	14.18	6.91	53.40	0.00	3.82	82.84	53.23	56.52	81.44	56.37	55.42	38.59	19.85	42.22	96.39	80.38	82.85	100.00	10.00	0.00
Impulsive transient by more than one point	59.50	23.95	43.47	94.36	17.66	57.47	83.79	49.20	68.25	81.75	55.27	73.43	68.02	16.94	77.51	95.97	81.19	79.46	100.00	12.35	100.00
Oscillatory transient	14.88	82.34	4.47	5.99	98.58	3.22	75.23	24.12	52.14	77.29	25.81	58.24	83.60	45.02	39.01	96.93	79.55	79.84	90.16	95.39	2.90
Interruption	25.81	19.14	48.67	18.64	14.60	56.59	84.29	88.81	33.22	81.34	88.20	33.20	95.69	100.00	31.72	91.04	80.26	79.98	96.36	86.74	58.19
Sag 0.3% to 0.5%	21.43	20.74	12.41	40.51	38.88	21.23	67.49	54.49	29.60	69.70	59.02	33.37	38.93	60.64	23.49	90.11	79.76	78.38	64.80	14.52	39.83
Sag 0.5% to 0.75%	36.48	43.36	38.86	44.83	46.79	47.71	53.03	42.57	47.49	58.31	35.85	45.67	33.42	28.32	41.62	89.91	79.61	79.08	87.70	88.14	59.29
Swell 1.25% to 1.5%	31.13	27.31	35.67	18.89	17.92	8.57	84.02	76.86	43.56	78.64	84.36	45.18	40.06	39.42	27.85	99.66	80.36	79.50	94.90	50.42	0.00
Sag + oscillatory	0.00	0.00	0.00	0.00	0.00	0.00	43.93	14.13	24.37	51.66	11.95	25.62	18.64	8.89	100.00	88.68	80.28	80.71	27.40	0.00	0.00
Healthy signal	0.00	0.00	0.00	0.00	0.00	0.00	80.83	12.07	78.50	74.91	25.20	63.00	99.98	0.00	51.00	100.00	95.28	100.00	96.35	0.00	0.00
Global	30.14	25.06	17.74	26.11	21.31	18.17	74.17	48.74	48.98	73.05	50.14	48.00	55.82	38.63	48.75	94.70	81.78	83.64	79.59	35.74	23.66

Number of simulations: 200. HOS_t, higher-order statistics in time domain; SK, spectral kurtosis.

ACKNOWLEDGEMENTS

The authors would like to thank the Spanish Government for funding the research Project *T EC*2010–19242–*C*03–03 (SIDER-HOSAPQ). This work is newly supported by the Spanish Ministry of Economy and Competitiveness in the frame of the Statal Plan of Excellency for Research, via the project *T EC*2013–47316–*C*3–2–*P* (SCEMS-AD-TED-PQR). Our unforgettable thanks to the trust we have from the Andalusian Government for funding the Research Group *P A I D I*–*T I C*–168 in Computational Instrumentation and Industrial Electronics (ICEI).

COMPETING INTERESTS

The authors declare that they have no competing interests.

REFERENCES

1. MHJ Bollen, S Bahramirad, A Khodaei, in *Proceedings on the 2014 IEEE 16th International Conference on Harmonics and Quality of Power (ICHQP)*. Is there a place for power quality in the smart grid? (University Politehnica of BucharestRomania, 2014), pp. 713–717.

2. EU Expert group 1:, Functionalities of smart gridsand smart meters. Technical report, EU Commission Task Force for Smart Grids (2009). http://ec.europa.eu/energy/en/topics/markets-and-consumers/smart-grids-and-meters.

3. Y Xiao, *Communication and networking in smart grids* (CRC Press, Broken Sound Parkway NW, Suite 300, 2012).

4. WK Reder, IEEE smart grid. Part 2: a grand vision for smart grid. Technical report. IEEE (2014). http://smartgrid.ieee.org/education.

5. MHJ Bollen, IY-H Gu, PGV Axelberg, E Styvaktakis, Classification of underlying causes of power quality disturbances: deterministic versus statistical methods. EURASIP J. Adv. Signal Process. 2007(1), 1–17 (2007).

6. YH Gu, MHJ Bollen, Time-frequency and time-scale domain analysis of voltage disturbances. IEEE Trans. Power Deliv. 15(4), 1279–1283 (2000).

7. MV Ribeiro, CAG Marques, CA Duque, AS Cerqueira, JLR Pereira, Detection of disturbances in voltage signals for power quality analysis using HOS. EURASIP J. Adv. Signal Process. 2007(1), 1–13 (2007).

8. MV Ribeiro, JLR Pereira, Classification of single and multiple disturbances in electric signals. EURASIP J. Adv. Signal Process. 2007(1), 1–18 (2007).

9. A Agüera-Pérez, JC Palomares-Salas, JJG de la Rosa, JM Sierra-Fernández, D Ayora-Sedeño, A Moreno-Muñoz, Characterization of electrical sags and swells using higher-order statistical estimators. Measurement. 44(Issue 8), 1453–1460 (2011).

10. ON Gerek, DG Ece, Power-quality event analysis using higher order cumulants and quadratic classifiers. IEEE Trans. Power Deliv. 21(2), 883–889 (2006).

11. ON Gerek, DG Ece, 2-D analysis and compression of power quality event data. IEEE Trans. Power Deliv. 19(2), 791–798 (2004).

12. DG Ece, ON Gerek, Power quality event detection using joint 2D wavelet subspaces. IEEE Trans. Instrumentation Meas. 53(4), 1040–1046 (2004).

13. O Poisson, P Rioual, M Meunier, Detection and measurement of power quality disturbances using wavelet transform. IEEE Trans. Power Deliv. 15(3), 1039–1044 (2000).

14. S Santoso, WM Grady, EJ Powers, J Lamoree, SC Bhatt, Characterization of distribution power quality events with fourier and wavelet transforms. IEEE Trans. Power Deliv. 15(1), 247–254 (2000).

15. JJG de la Rosa, JM Sierra-Fernández, A Agüera-Pérez, JC Palomares-Salas, A Moreno-Muñoz, An application of the spectral kurtosis to characterize power quality events. Electrical Power Energy Syst. 49, 386–398 (2013).

16. JJG de la Rosa, JM Sierra-Fernández, A Agüera-Pérez, JC Palomares-Salas, A Jiménez-Montero, A Moreno-Muñoz, in*Proceedings on the IEEE International Workshop on the Applied Measurements for Power Systems (AMPS), vol. 1*. Power quality events' measurement criteria based in higher-order statistics: towards new measurement indices (IEEEAachen, Germany, 2013), pp. 73–79.

17. JJG de la Rosa, A Agüera-Pérez, JC Palomares-Salas, JM Sierra-Fernández, A Moreno-Muñoz, A novel virtual instrument for power quality surveillance based in higher-order statistics and case-based reasoning. Measurement. 45(7), 1824–1835 (2012).

18. JC Palomares-Salas, JJG de la Rosa, A Agüera-Pérez, A Moreno-Muñoz, Intelligent methods for characterization of electrical power quality signals using higher order statistical features. Przeglad Elektrotechniczny. 2012(8), 236–243 (2012).

19. S Haykin, *Neural Networks* (Englewood Cliffs, NJ, 1994).

20. DD Ferreira, CAG Marques, JM de Seixas, AS Cerqueira, MV Ribeiro, CA Duque, *Exploiting Higher-Order Statistics Information for Power Quality Monitoring* (InTech Open Science, 2011). http://intechopen.com.

21. L Angrisani, P Daponte, M D'Apuzzo, Wavelet network-based detection and classification of transients. IEEE Trans. Instrumentation Meas. 50(5), 1425–1435 (2001).

22. I Monedero, C León, J Ropero, A García, JM Elena, JC Montaño, Classification of electrical disturbances in real time using neural networks. IEEE Trans. Power Deliv. 22(3), 1288–1296 (2007).

23. M Valtierra-Rodriguez, de Jesus Romero-Troncoso R, RA Osornio-Rios, A Garcia-Perez, Detection and classification of single and combined power quality disturbances using neural networks. IEEE Trans. Ind. Electron. 61(1), 2473–2482 (2014).

24. DD Ferreira, AS Cerqueira, CA Duque, MV Ribeiro, HOS-based method for classification of power quality disturbances. Electron. Lett. 45(3), 183–185 (2009).

25. JJG de la Rosa, A Agüera-Pérez, JC Palomares-Salas, A Moreno-Muñoz, Higher-order statistics: discussion and interpretation. Measurement. 46(8), 2816–2827 (2013).

26. Z Liu, Q Zhang, Z Han, G Chen, A new classification method for transient power quality combining spectral kurtosis with neural network. Neurocomputing. 125(1), 95–101 (2014).

27. S Mallat, *A wavelet tour of signal processing* (Academic Press, Burlington, MA 01803, 2009).

28. J Sola, J Sevilla, Importance of input data normalization for the application of neural networks to complex industrial problems. IEEE Transa. Nuclear Sci. 44(3), 1464–1468 (1997).

29. JM Mendel, Tutorial on higher-order statistics (spectra) in signal processing and system theory: theoretical results and some applications. Proc. IEEE. 79(3), 278–305 (1991).

30. CL Nikias, JM Mendel, Signal processing with higher-order spectra. IEEE Signal Process. Mag., 10–37 (1993).

31. CL Nikias, AP Petropulu, *Higher-Order Spectra Analysis. A Non-Linear Signal Processing Framework* (Prentice-Hall, Englewood Cliffs, NJ, 1993).

32. AK Nandi, *Blind Estimation Using Higher-Order Statistics, vol 1,* 1st edn. (Kluwer Academic Publishers, Boston, 1999).

33. JJG de la Rosa, A Moreno-Muñoz, Electrical transients monitoring via higher-order cumulants and competitive layers. Przeglad Elektrotechniczny - Electrical Rev. 85(10), 284–289 (2009).

34. J Jakubowski, K Kwiatos, A Chwaleba, S Osowski, Higher order statistics and neural network for tremor recognition. IEEE Trans. Biomed. Eng. 49(2), 152–159 (2002).

35. JJG de la Rosa, I Lloret, CG Puntonet, JM Górriz, Higher-order statistics to detect and characterize termite emissions. Electron. Lett. 40(20), 1316–1317 (2004).

Chapter 6

Probabilistic Assessment of Power System Performance Quality

Badr M. Alshammari, Mohamed A. El-Kady

SEC Chair in Power System Reliability and Security, King Saud University, Riyadh, KSA

ABSTRACT

In a recently published work by the authors, a novel framework was developed and applied for assessment of reliability and quality performance levels in real-life power systems with practical large-scale sizes. The new assessment methodology is based on three metaphors (dimensions) representing the relationship between available generation capacities and required demand levels. The developed reliability and performance quality indices were deterministic in nature. That is, they represent one operating state (a snapshot of the system conditions) in which the required demand as well as the generation and transmission capacities are known with 100% certainty. In real life, however, load variations occur randomly so as the contingencies which cause some generation and/or transmission capacities to be lost (become unavailable). In other words, neither the load levels nor the generation or transmission capacities are known with absolute certainty. They are rather subject to random variations and, consequently, the calculated reliability and performance quality indices are all subject to random variations where only expected values of these indices can be evaluated. This paper presents a major extension to the previously published work by developing a theory and formulas for computing the expected values of different system reliability and performance quality indices. In this context, a "contingency scenario" or a system "demand level" are regarded, in a more general sense, as a "state", which occurs with certain probability and represents a given demand value and availability pattern of various capacities in the system. The work of this paper provides a practical and meaningful methodology for real-life assessment of power system reliability and performance quality levels. Practical applications are also presented, for demonstration purposes, to the Saudi electricity power grid.

KEYWORDS

Power Systems; Probabilistic Analysis; Reliability; Quality Assessment; Linear Programming

1. INTRODUCTION

Maintaining a continuous and sufficient power supply to the customers at a reasonable cost is the prime objective of electric power companies around the world. In this regard, power system cost-effectiveness, security, adequacy and reliability analyses have become a major concern in today's highly-competitive business environment of power utility planning and operations [1-3]. In a recent paper by the authors [4], a novel framework was developed and applied for assessment of reliability and quality performance levels in real-life power systems with practical largescale sizes. The new assessment methodology is based on three metaphors (dimensions) representing the relationship between available generation capacities and required demand levels. The first metaphor defines whether or not the capacity exists, the second metaphor defines whether or not the capacity is needed, and the last metaphor defines whether or not the capacity can reach (delivered to) the demand. The eight possible combinations associated with the 0/1 (Yes/No) values of the three metaphors would, in turn, define a set of powerful systemwide performance quality measures relating to generation deficiency, redundancy, bottling, etc. The developed reliability and performance quality indices were deterministic in nature. That is, they represent one operating state (a snapshot of the system conditions) in which the required demand as well as the generation and transmission capacities are known with 100% certainty.

In real life, however, load variations occur randomly so as the contingencies which cause some generation and/or transmission capacities to be lost (become unavailable). In other words, neither the load levels nor the generation or transmission capacities are known with absolute certainty. They are rather subject to random variations and, consequently, the calculated reliability and performance quality indices are all subject to random variations where only expected values of these indices can be evaluated.

Methods for computing probabilistic contingency-based reliability and performance quality indices have previously been published in the literature [5-7]. These methods are based on a combined contingency analysis and reliability evaluation scheme which integrates both the contingency effect and its probability of occurrence into one routine of analysis. In the present research work, similar analysis will be used to compute the expected values of different system reliability and performance quality indices. In this context, a "contingency scenario" or a system "demand level" are regarded, in a more general sense, as a "state", which occurs with certain probability and represents a given demand value and availability pattern of various capacities in the system.

The work of this paper presents a major extension to the previously published work [4] by developing a theory and formulas for computing the expected values of different system reliability and performance quality indices. In this context, a "contingency scenario" or a system "demand level" are regarded, in a more general sense, as a "state", which occurs with certain probability and represents a given demand value and availability pattern of various capacities in the system.

The work of this paper provides a practical and meaningful methodology for real-life assessment of power system reliability and performance quality levels. Practical applications are also resented in the Saudi electricity power grid.

2. POWER SYSTEM QUALITY ASSESSMENT

2.1. Performance Quality Framework

In the framework presented in [4], three metaphors (dimensions) were introduced to represent the relationship between certain system generation capacity and the demand. These metaphors relate to the following demand fulfillment issues:

1) Need of capacity for demand fulfillment;

2) Existence of capacity (availability for demand fulfillment);

3) Ability of capacity to reach the demand.

The first metaphor defines whether or not the capacity is needed, the second metaphor defines whether or not the capacity exists, and the last metaphor defines whether or not the capacity can reach (delivered to) the demand. The eight possible combinations associated with the 0/1 (Yes/No) values of the three metaphors would, in turn, define a set of powerful system-wide performance quality measures, namely:

1) Utilized: A given capacity is said to be utilized if it is needed (for demand fulfillment), exists, and can reach the demand;

2) Bottled: A given capacity is said to be bottled if it is needed (for demand fulfillment) and exists, but cannot reach the demand;

3) Shortfall: A given capacity is said to be shortfall if it is needed (for demand fulfillment) and, anyhow, does not exist and can reach the demand;

4) Deficit: A given capacity is said to be deficit if it is needed (for demand fulfillment) but, however, does not exist and cannot reach the demand;

5) Surplus: A given capacity is said to be surplus if it is not needed (for demand fulfillment) although exists and can reach the demand;

6) Redundant: A given capacity is said to be redundant if it is not needed (for Demand fulfillment) although exists but, anyhow, cannot reach the demand;

7) Spared: A given capacity is said to be spared if it is not needed (for demand fulfillment) and, anyhow, does not exist although can reach the demand;

8) Saved: A given capacity is said to be saved if it is no needed (for demand fulfillment) and, anyhow, does not exist and cannot reach the demand.

We note here that the above performance quality measures are associated with different combinations (topples) of the three quality metaphors, namely, "existence", "need" and "ability to reach the demand". The corresponding quality state of a given capacity can be represented by a threevalue expression of either a "Yes/No" or "1/0" type indicating the true/false value associated with each quality metaphor.

The evaluation of the above quality indices requires the knowledge of the following data types for the demand and various system facilities:

1) The value of demand required to be supplied;

2) The value of generation capacity as well as the maximum site capacity (the limit of potential increase in existing generation capacity);

3) The value of transmission capacity.

2.2. Linear Program Formulation

In the computational scheme of [4], the integrated system quality assessment is performed via solving a master linear programming problem [8] in which a feasible power flow is established which minimizes the total system non-served load subject to capacity limits and flow equations. The master linear program, which utilizes the network bus incidence matrix A, is formulated as

$$\left.\begin{array}{l} \text{Minimize } f = \sum_{l=1}^{nL}(-P_l) \text{ with respect to} \\ \mathbf{P}_L, \mathbf{P}_G \text{ and } \mathbf{P}_T \text{ such that} \\ \qquad \mathbf{A}\,\mathbf{P}_T = \begin{bmatrix} -\mathbf{P}_L \\ \mathbf{P}_G \end{bmatrix} \\ \qquad \mathbf{P}_L \le \overline{\mathbf{P}}_L,\ -\mathbf{P}_L \le \mathbf{0} \\ \qquad \mathbf{P}_G \le \overline{\mathbf{P}}_G,\ -\mathbf{P}_G \le \mathbf{0} \\ \qquad \mathbf{P}_T \le \overline{\mathbf{P}}_T,\ -\mathbf{P}_T \le \overline{\mathbf{P}}_T \end{array}\right\} \tag{1}$$

In the above master linear program,

$\overline{\mathbf{P}}_T$ = vector of n_T elements representing transmission branch capacities;

$\overline{\mathbf{P}}_L$ = vector of n_L elements of peak bus loads;

$\overline{\mathbf{P}}_G$ = vector of n_G elements representing generator capacities.

Also, in the above master linear program (1), P_L, P_G, and P_T are n_L, n_G and n_T column vectors representing the actual load bus powers (measured outward), generator bus powers (measured inwards) and transmission line powers (measured as per the network bus incidence matrix A), respectively. The solution of the above linear program provides a more realistic (less conservative) flow pattern in view of the fact that when load curtailments are anticipated, all system generation resources would be re-dispatched in such a way which minimizes such load cuts. The feasible flow pattern established from the Master Linear Program is then used to evaluate various integrated system quality indices through a set of closely related sub-problems.

2.3. Implementation Mechanisms

For real life power systems with practical sizes, the quality indices cannot be evaluated by inspection. An appropriate computerized scheme is needed in order to properly evaluate various quality indices according to their stated definitions. The master linear program presented before forms the bases for analyzing and evaluating the quality indices. For example, the Load Supply Reliability can be evaluated as follows:

$$LNS_l = \text{Load Not} - \text{Served at Load Bus } (l) = \left(\overline{P}_l - P_l^{(1)}\right)$$

$$LNS = \text{Total System Load Not} - \text{Served} = \sum_{l=1}^{nL}\left(\overline{P}_l - P_l^{(1)}\right)$$

where the bus loads at the solution of the master linear program are termed as $P_l^{(1)}$, and P_l denotes the solution load value at bus (l).

On the other hand, generation quality indices are defined in terms of the previously defined "1/0" states indicating the (Needed, Exists, Can-reach) true/false values associated with each quality metaphor. We shall use the symbol $Q_g ijk$ to indicate the generation quality index state. Also, in the following expressions, we shall use Min{x, y, ···, z} to indicate the minimum of x, y, ···, z. The notation will be used to denote Max{0, x}, that is the maximum of x and zero (=x if x > 0, or 0 otherwise). For example, the Utilized Generation Capacity index is given by

$$Q_g 111 = \text{Utilized Capacity} \equiv \{\text{needed, exists, can reach}\}$$
$$= \sum_{l=1}^{nL} \left(P_l^{(1)}\right)$$

Similarly, the Bottled Generation Capacity index is given by

$$Q_g 110$$
$$= \text{Bottled Capacity} \equiv \{\text{needed, exists, cannot reach}\}$$
$$= \text{Min}\left\{\left[\sum_{l=1}^{nL} \bar{P}_l - \sum_{g=1}^{nG} P_g^{(1)}\right], \left[\sum_{g=1}^{nG} \text{Max}\left\{0, \left(\bar{P}_g - P_g^{(1)}\right)\right\}\right]\right\}$$

Also, the Surplus Generation Capacity ($Q_g 011$) is calculated as

$$Q_g 011$$
$$= \text{Surplus Capacity} \equiv \{\text{notneeded, exists,can reach}\}$$
$$= \text{Min}\left\{\left[\text{Max}\left\{0, \left(\sum_{g=1}^{nG} \bar{P}_g - \sum_{l=1}^{nL} \bar{P}_l\right)\right\}\right],\right.$$
$$\left.\left[\text{Max}\left\{0, \left(\sum_{g=1}^{nG} P_g - \sum_{l=1}^{nL} \bar{P}_l\right)\right\}\right]\right\}$$

where the generation output values P_g are calculated at the solution of the linear program with open limits on the loads.

3. PROBABILISTIC ASSESSMENT

3.1. Probabilistic Reliability Indices

The power system can be described, for the purpose of composite reliability and performance quality assessment, by the three-component model, as shown in **Figure 1**, in which generation, transmission and load are considered as multi-state elements of the power system.

For a given operating state m, the values of the network variables will be the solution of the maximum load-supply optimization problem described in the previous section. Also, let f_m be the probability of operating state m (the sum of f_m for all m, including base-case scenario is 1). Then, the following three system-wide reliability indices may be defined:

1) Loss of Load Probability

$$LOLP^{(m)} = \sum_{m=1}^{M_s} LOLP^{(m)} \tag{2a}$$

Where

$$LOLP^{(m)} \equiv \mathrm{Max}_{\ell} \left\{ LOLP_{\ell}^{(m)} \right\} \tag{2b}$$

represents the system loss of load probability for any operating state m (load level, loss of generation and/or transmission capacities) in the power grid,

$$LOLP_{\ell}^{(m)} \equiv \lambda_{\ell}^{(m)} f_m \tag{2c}$$

represents the loss of load probability at bus l for operating state m,

$$\lambda_\ell^m \equiv \begin{cases} 0 & \text{if} \quad P_\ell^{(m)} \le P_\ell^o \\ 1 & \text{if} \quad P_\ell^{(m)} > P_\ell^o \end{cases} \tag{2d}$$

and P_ℓ^o denotes the scheduled (required) load at load bus l. Also, in Equation (3.8), M_s denotes the number of all possible states.

Figure 1. System model reliability evaluation.

2) Expected Load Not-Served

$$e(LNS) \equiv \sum_{\ell=1}^{nL} (eLNS_\ell) \tag{3a}$$

where n_L is the number of load buses in the system,

$$(eLNS_\ell) \equiv \sum_{m=1}^{M_c} \left(eLNS_\ell^{(m)}\right) \tag{3b}$$

represents the expected value of load not-served at bus l,

$$\left(eLNS_\ell^{(m)}\right) \equiv f_m LNS_\ell^{(m)} \tag{3c}$$

represents the expected value of Load Not-Served at bus l for the operating state m, and

$LNS_\ell^{(m)}$ = Load not served at buslfor operating state m

which is obtained from the solution of the Master Linear Program (1).

3) Expected Energy Not Served

$$(eENS) \equiv \sum_{\ell=1}^{nL} (eENS_\ell) \tag{4a}$$

Where

$$(eENS_\ell) \equiv \sum_{m=1}^{M_c} \left(eENS_\ell^{(m)}\right) \tag{4b}$$

$$\left(eENS_\ell^{(m)}\right) \equiv f_m ENS_\ell^{(m)} \tag{4c}$$

represents the expected value of energy not served at a bus l, and

$$\left(eENS_\ell^{(m)}\right) \equiv f_m ENS_\ell^{(m)} \tag{4d}$$

represents the expected value of energy not served at bus l for operating state m,

$$ENS_\ell^{(m)} \equiv T^{(m)} LNS_\ell^{(m)} \tag{4e}$$

represents the energy not served at bus l for operating state m, and $T^{(m)}$ denotes the time duration of operating state m.

3.2. Probabilistic Performance Quality Indices

Probabilistic performance quality indices can be calculated using the previously derived formulas based on the solution of the master linear program (1) subject to random variations of system demand level as well as forced outages in various generation and transmission facilities.

For example, the load variations, which are accounted for using the so-called "load-duration curves" can be used to calculate the expected value of the Load Not-Served (LNS), which is widely known as the Expected Load NotServed (eLNS).

On the other hand, the randomness in the generation and transmission capacity availability are accounted for using the so-called forced-outage rates (or availability rates) associated with various facilities. Consequently, the expected values of the performance quality indices Q_g111, Q_g110, Q_g101, etc., denoted by eQ_g111, eQ_g110, eQ_g101, etc., can be evaluated using the modeled randomness of the system load as well as the generation and transmission capacity availabilities.

4. APPLICATIONS TO SEC POWER SYSTEM

4.1. SEC Quality Performance Indices

In a recently completed industry supported study, applications were conducted on a practical power system comprising a portion of the interconnected Saudi power grid. The power system consists of two main regions, namely the Central region and the Eastern region. The two systems are interconnected through two 380 kV and one 230 kV double-circuit lines. Four zones are identified in the present analysis, three in the Central region (Riyadh, Qassim and Hail zones) and one in the Eastern region. In this application, three reliability and quality performance indices are considered, namely the system Load Not-Served (LNS), Utilized Generation Capacity (Q_g111) and the Bottled Generation Capacity (Q_g110). In the present work, a particular focus will be on Qassim and Hail zones for demonstration purposes. The system models used for these two zones are shown in Figures 2(a) and (b), respectively. **Table 1** outlines the network data in terms of generation and transmission facilities as well as system loads. Figures 3(a) and (b), on the other hand, summarize the results of the performance quality measures applied to the SEC power system for various system status (isolated or connected) of each zone.

In particular, Figures 3(a) and (b) depict the variation of the variation of quality indices (LNS, Q_g111 and Q_g110) with the required load level of the Qassim isolated and interconnected network, respectively.

Figures 4(a) and (b), on the other hand, show 3-dimensional graphs depicting the variation of Utilized Generation Capacity index Q_g111 with both load and generation capacity levels of the Hail isolated and interconnected network, respectively.

The results obtained reveal several important observations. For example, the results obtained for the isolated network scenario of Qassim zone (**Figure 3**(a)) show that the Load Not-Served is non-zero even for relatively low demand levels as it increases continuously from 300 MW at a demand level of 1840 MW to reach 2400 MW when the demand level is 4410 MW. This problem is clearly mitigated in the interconnected network scenario of Qassim zone (**Figure 3**(b)), where generation support from Riyadh zone becomes available. In this case, the Load NotServed stays at zero value for all demand levels up to 2620 MW where it starts to increase slowly to reach 70 MW at a demand level of 3370 MW before it starts to increase sharply afterwards to reach about 2000 MW at a demand level of 5610 MW.

The Utilized Generation Capacity index Q_g111 for the isolated network scenario of Qassim zone (**Figure 3**(a)) increases continuously with the required demand level until it saturates at about 2000 MW when the required demand reaches 2943 MW when no more available generation can be utilized. This situation is avoided—as expected— in the interconnected network scenario of Qassim zone (**Figure 3**(b)) where the Utilized Generation Capacity increases continuously to reach, for example, 3600 MW at demand level of 5610 MW as more generation support becomes available. The Bottled Generation Capacity index Q_g110, for the isolated network scenario of Qassim zone (**Figure 3**(a)), decreases continuously with the required demand until it disappears at a demand level of 2575 MW. In the case of the interconnected network scenario of Qassim zone (**Figure 3**(b)), however, the Bottled Generation Capacity coincides with the Load Not-Severed for all required demand levels up to 4115 MW. After this level, the Bottled Generation Capacity starts to decrease continuously.

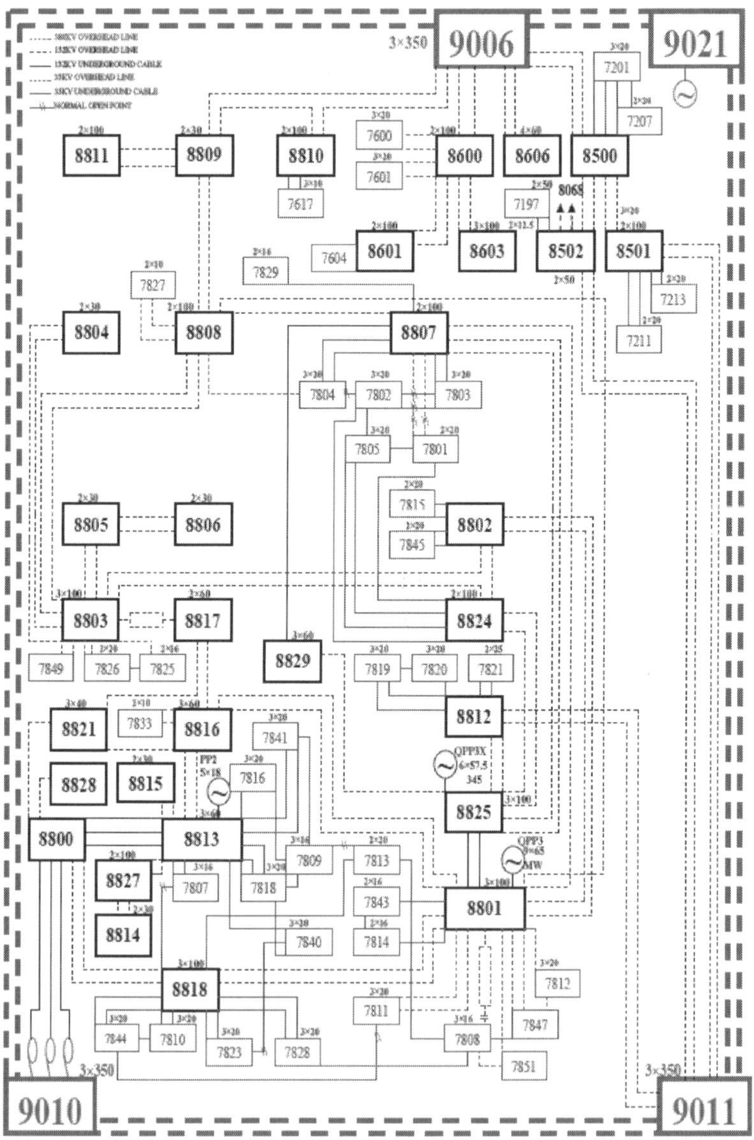

(a)

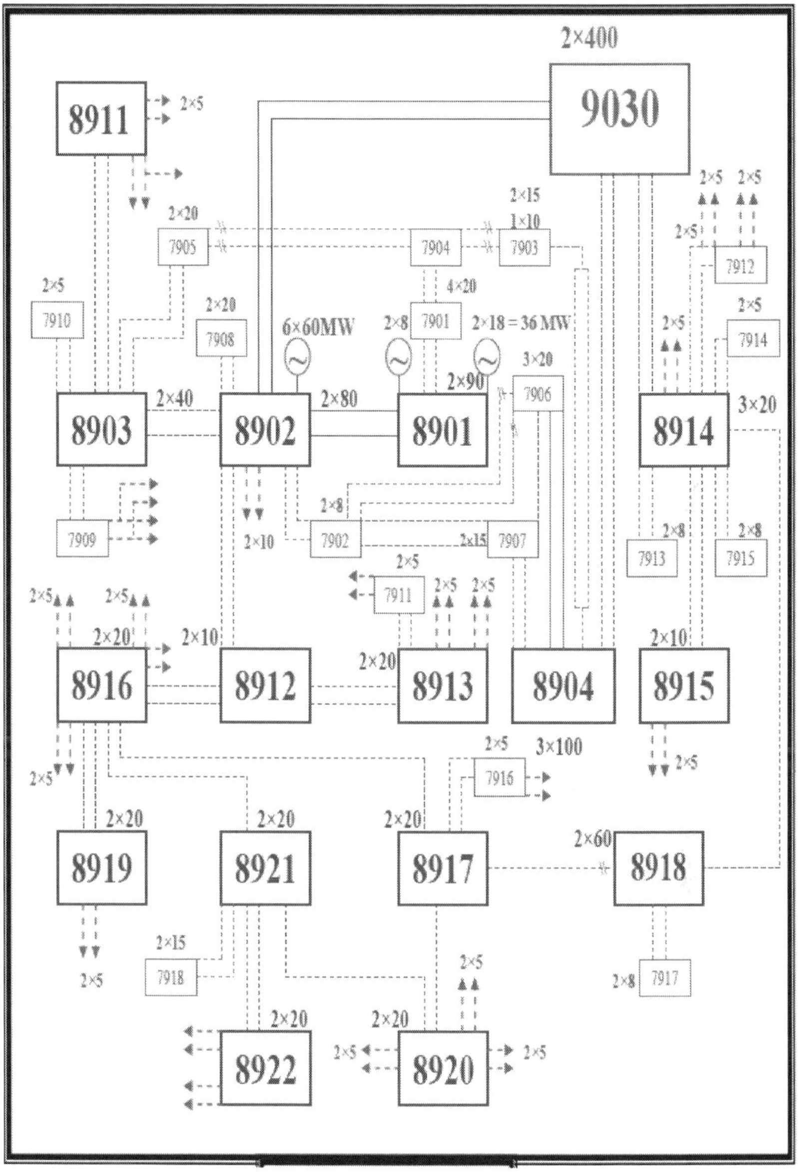

(b)

Figure 2. (a) Single-line diagram of SEC—qassim zone; (b) Single-line diagram of SEC—hail zone.

Table 1. Generation, transmission and loads of SEC power system.

Network	State	Generators		Transmissions	Loads	
		Value	Number	Number	Value	Number
Hail	Isolated	593.6395	9	68	655.3934	46
	Interconnected	1393.6395	10	69	655.3934	46
Qassim	Isolated	2008.0355	21	116	3679.4002	76
	Interconnected	4108.0355	24	121	3741.1541	78

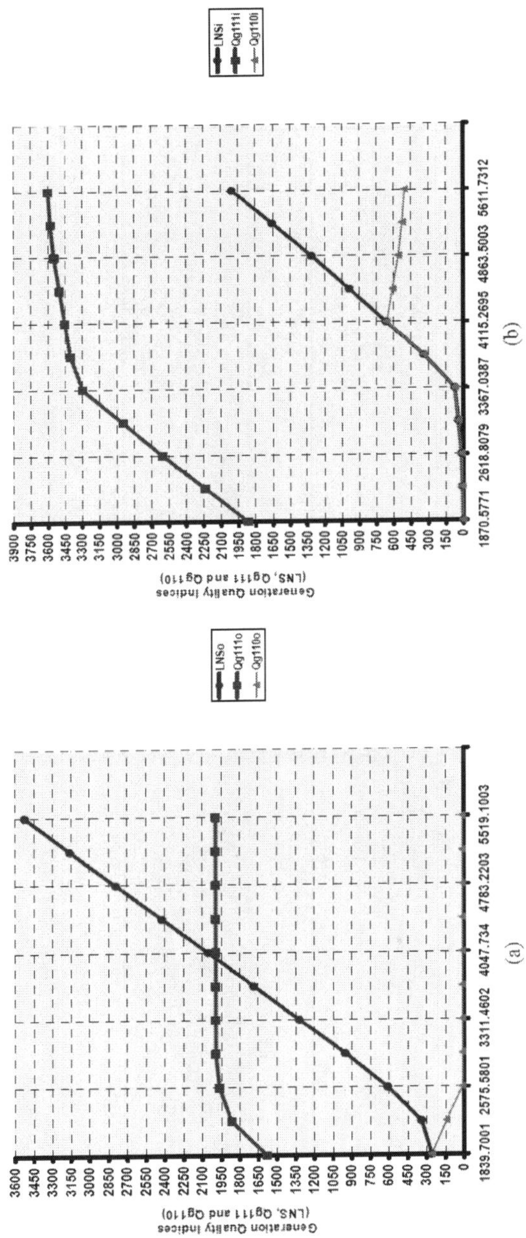

Figure 3. (a) Variation of quality indices (LNS, Q_g111 and Q_g110) with the required load level of the qassim isolated network; (b) Variation of quality Indices (LNS, Q_g111 and Q_g110) with the required load level of the qassim interconnected network.

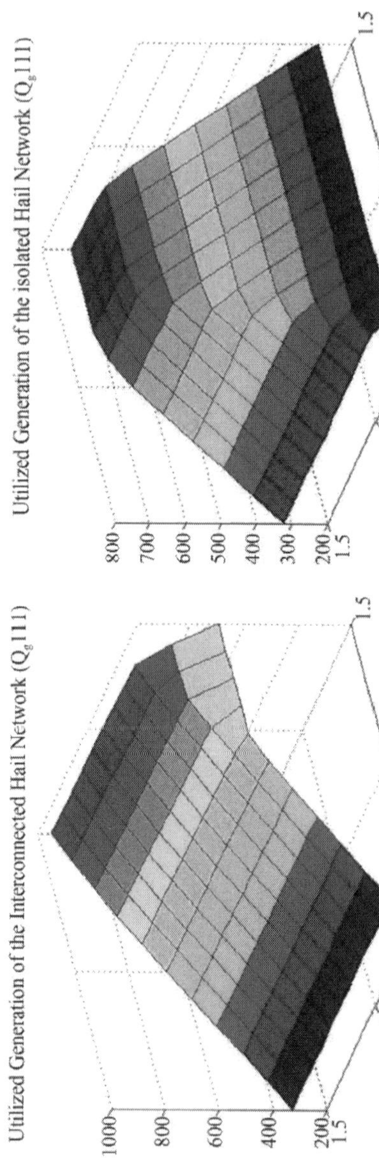

Figure 4. (a) 3-dimensional graph showing variation of utilized generation index Q_g111 with both load and generation capacity levels of the Hail isolated network; (b) 3-dimensional graph showing variation of utilized generation index Q_g111 with both load and generation capacity levels of the Hail interconnected network.

4.2. Probabilistic Analysis of Hail System

As was stated before, the Hail network under investigation contains 10 generators, 69 branches (transmission lines, underground cables and power transformers) as well as 46 loads. In order to simplify the probabilistic assessment, the total available generation is represented by three equivalent units, two of which represent the internal available generation within Hail and the third represents the interconnection support from Qassim. The availability rate of the generating unit is 0.9825. On the other hand, only the five major transmission lines in Hail are considered in the probabilistic assessment with equal availability rate of 0.985. The other transmission lines are assumed to be available all the time.

Based on the actual load duration curve of Hail, shown in **Figure 5**, the system load is assumed to have seven possible levels, namely 150 MW, 200 MW, 300 MW, 400 MW, 500 MW, 600 MW and 650 MW with probabilities of occurrence (calculated from the load duration curve) equal 0.009, 0.2, 0.314, 0.182, 0.045, 0.23 and 0.023, respectively.

Using the results of the probabilistic analysis of Hail system, the discrete probability density functions of various reliability and performance quality indices can be evaluated and displayed. These discrete density functions show the overall probabilities of occurrence associated with certain values of the system performance indices. The probability density function of the Load Not Served (LNS) for Hail network at required load level 150 MW is depicted in **Figure 6**(a). Also, **Figure 6**(b) shows the probability density of the Surplus Generation Capacity (Q_g011) at load level 200 MW.

Some valuable information can be drawn from the probability density functions of Figures 6(a) and (b). For example, the probability of Load Not Served (LNS) in Hail being zero when the required load level is 150 MW is 0.9387 (from **Figure 6**(a)) while there is a probability of 0.0310 that the unsupplied load is equal to, or greater than 25 MW. On the other hand, there is a probability of 0.92 that the Surplus Generation

Capacity (Q_g011) in Hail (at load level 200 MW) is equal to, or greater than 1175 MW.

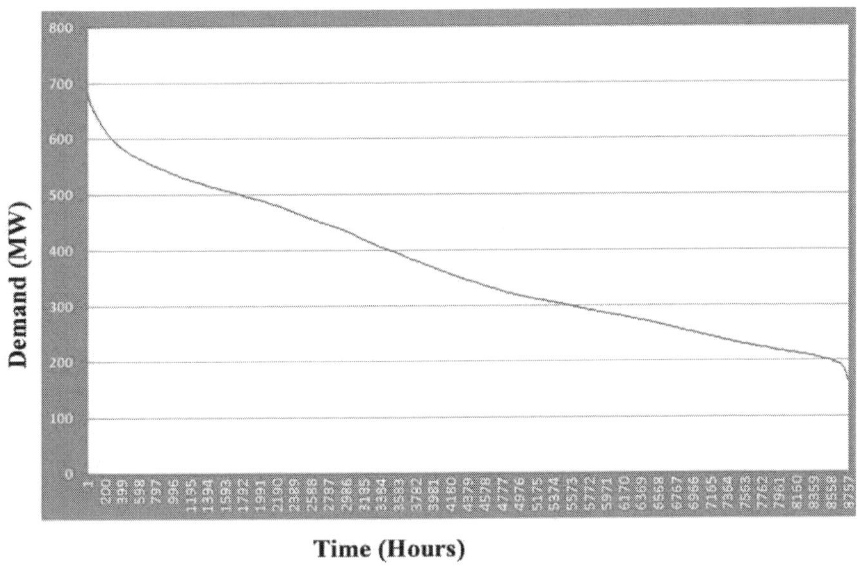

Figure 5. Hail load duration curve.

When all required load levels are considered with their respective probabilities of occurrence (as per the load duration curve), an overall estimate of the expected value of reliability and performance quality indices can be obtained as shown in **Table 2**.

The overall expected value of the Load Not Served (LNS) for Hail is 4.01 MW, which is less than 1% of the required load in Hail. On the other hand, overall expected value of the Bottled Generation Capacity (Q_g110) is 3.86 (MW), which again is less than 1% of the total Hail generation. It is also of interest to note that the overall expected value of the Utilized Generation Capacity (Q_g111) is 378.1 (MW), which represents almost 64% of the total generation capacity in Hail. In other words, about 36% of Hail available generation capacity is expected to be unutilized. Recalling from **Table 1** that the Hail generation capacity is 594 MW (base value) while the required base load is 655 MW, one may conclude that the 64% expected value for the utilized generation is a reflection of the relatively large variations of the required load during different periods of the year, as is evident from the load duration curve of**Figure 5**.

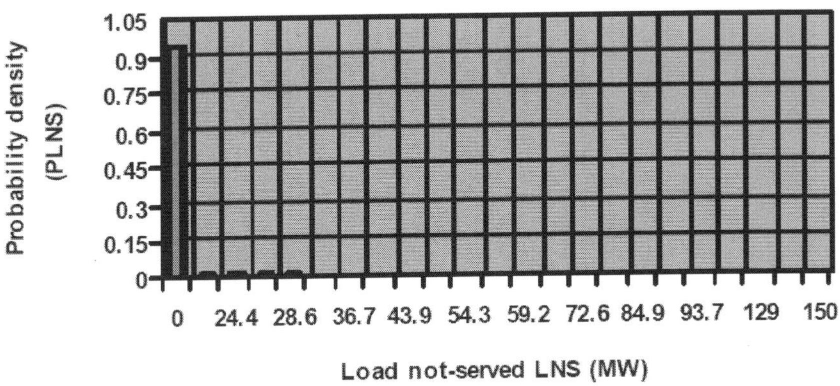

(a)

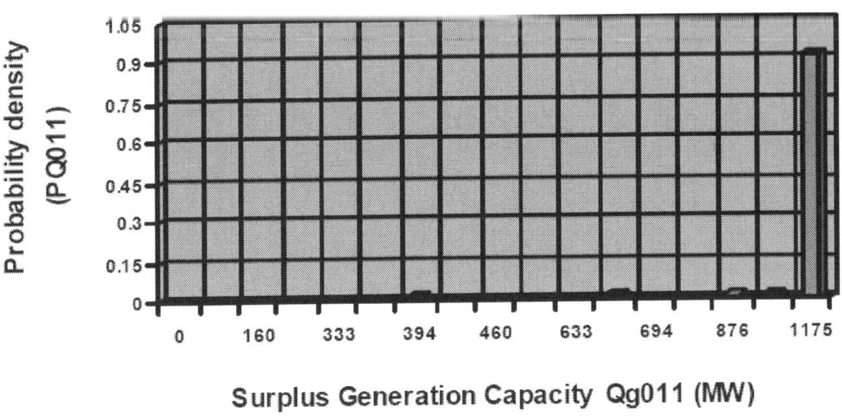

(b)

Figure 6. (a) Probability density of load not served (LNS) for Hail network at load 150 MW; (b) Probability density of surplus generation capacity (Q_g011) for Hail network at load 200 MW.

Table 2. Expedited values of reliability and performance quality indices for Hail network.

Index	Expected Value
Load Not Served (*LNS*)	4.01 (MW)
Utilized Generation Capacity (Q_g111)	378.1 (MW)
Bottled Generation Capacity (Q_g110)	3.86 (MW)
Surplus Generation Capacity (Q_g011)	962.5 (MW)
Redundant Generation Capacity (Q_g010)	24.8 (MW)

5. CONCLUSIONS

The work of this paper represents a major extension to the previously published work by developing a theory and formulas for computing the expected values of different system reliability and performance quality indices. The reliability and performance quality indices, when evaluated at a given load level and a certain scenario of available generation and transmission capacities, would provide indication on system performance for only such a particular system condition (snapshot). However, the novel formulation presented in this paper can accommodate the randomness associated with the load level as well as the availability of generation and transmission capacities. In this case, expected values of reliability indices, such as the Expected Load Not-Served (eLNS), as well as the expected values of performance quality indices, such as Utilized Generation Capacity (eQ_g111), Bottled Generation Capacity (eQ_g110), Shortfall Generation Capacity (eQ_g101), Deficit Generation Capacity (eQ_g100), Surplus Generation Capacity (eQ_g011), Redundant Generation Capacity (eQ_g010), Spared Generation Capacity (eQ_g001) and Saved Generation Capacity (eQ_g000) can be calculated using the load duration curve as well as the availability rates of generation and transmission facilities in the system.

The results of the probabilistic analysis of Hail system have revealed that the probability of Load Not Served (LNS) in Hail being zero (when the required load level is 150 MW) is 0.9387, while there is a probability of 0.0310 that the unsupplied load is equal to, or greater than 25 MW. When all required load levels are considered with their respective probabilities of occurrence (as per the load duration curve), an overall estimate of the expected value of reliability and performance quality indices were obtained for Hail where the overall expected value of the Load Not Served (LNS) is 4.01 MW, which is less than 1% of the required load in Hail. On the other hand, overall expected value of the Bottled Generation Capacity (Q_g110) is 3.86 (MW), which again is less than 1% of the total Hail generation.

ACKNOWLEDGEMENTS

This work was supported by the Saudi Electricity Company.

REFERENCES

1. H. Jun, Y. Zhao, P. Lindsay and P. Kit, "Flexible Transmission Expansion Planning with Uncertainties in an Electricity Market," IEEE Transactions on Power Systems, Vol. 24, No. 1, 2009, pp. 479-488.

2. M. El-Kady, M. El-Sobki and N. Sinha, "Reliability Evaluation for Optimally Operated Large Electric Power Systems," IEEE Transactions on Reliability, Vol. R35, No. 1, 1986, pp. 41-47.

3. M. El-Kady, M. El-Sobki and N. Sinha, "Loss of Load Probability Evaluation Based on Real Time Emergency Dispatch," Canadian Electrical Engineering Journal, Vol. 10, 1985, pp. 57-61.

4. M. A. El-Kady and B. M. Alshammary, "A Practical Framework for Reliability and Quality Assessment of Power Systems," Journal of

Energy and Power Engineering (EPE), Vol. 3, No. 4, 2011, pp. 499-507.

5. P. Jirutitijaroen and C. Singh, "Reliability Constrained Multi-Area Adequacy Planning Using Stochastic Programming With Sample-Average Approximations," IEEE Transactions on Power Systems, Vol. 23, No. 2, 2008, pp. 405-513.

6. R. Billinton and D. Huang, "Effects of Load Forecast Uncertainty on Bulk Electric System Reliability Evaluation," IEEE Transactions on Power Systems, Vol. 23, No. 2, 2008, pp. 418-425.

7. M. El-Kady, B. Alaskar, A. Shaalan and B. Al-Shammri, "Composite Reliability and Quality Assessment of Interconnected Power Systems," International Journal for Computation and Mathematic in Electrical and Electronic Engineering (COMPEL), Vol. 26, No. 1, 2007.

8. P. Gill, W. Murray and M. Wright, "Practical Optimization," Academic Press, Waltham, 1981.

Chapter 7

Voltage sag compensation strategy for unified power quality conditioner with simultaneous reactive power injection

Yunfei XU1, Xiangning XIAO1, Yamin SUN1, Yunbo LONG1

1 State Key Laboratory of Alternate Electrical Power System with Renewable Energy Sources, North China Electric Power University, Beijing 102206, China

ABSTRACT

Unified power quality conditioner (UPQC) holds the capability of solving power quality problems, especially shows good performance in the voltage sag compensation. In this paper, a compensation strategy based on simultaneous reactive power injection for UPQC (namely UPQC-SRI) is proposed to address the issue of voltage sag. The proposed UPQC-SRI determines the injection angle of compensation voltage with consideration of optimal configuration of UPQC current-carrying. Moreover, the compensation strategy also considers the current-carrying limit of UPQC, and then the zero active power injection region of UPQC-SRI (also called UPQC-SRI region) is obtained. Under the conditions which exceed the UPQC-SRI region, the limit value of shunt current is determined by this proposed strategy. Finally, the proposed strategy and the corresponding algorithm are verified under the PSCAD/EMTDC platform. The result indicates the proposed UPQC-SRI strategy in this paper can provide more persistent voltage sag compensation than the previous strategies for the sensitive load.

KEYWORDS

Unified power quality conditioner (UPQC), Voltage sag, Simultaneous reactive power injection, Zero active power injection compensation region

1. INTRODUCTION

As the non-linear power load increases and the structure of distribution grid is more complex, the problems of power quality become more and more serious [1]. Especially, voltage quality problem could severely affect the normal operation of the sensitive load largely connected to the distribution grid, which leads to giant economic losses and negative effects. The complex and coupled power quality problems are commonly widespread in industries, such as automobile and electronics manufacturing, hospital and entertainment facilities [2, 3]. As the comprehensive compensation equipment of power quality, unified power quality conditioner (UPQC) is especially appropriate for solving the problems above, which can mitigate both voltage and current quality issues such as harmonic, unbalance, voltage swell and sag [4]. As shown in Fig. 1, the two converters of UPQC with the back to back combination are connected to power grid in series and shunt respectively, which makes it more suitable for voltage sag compensation with internal energy exchange [5–7].

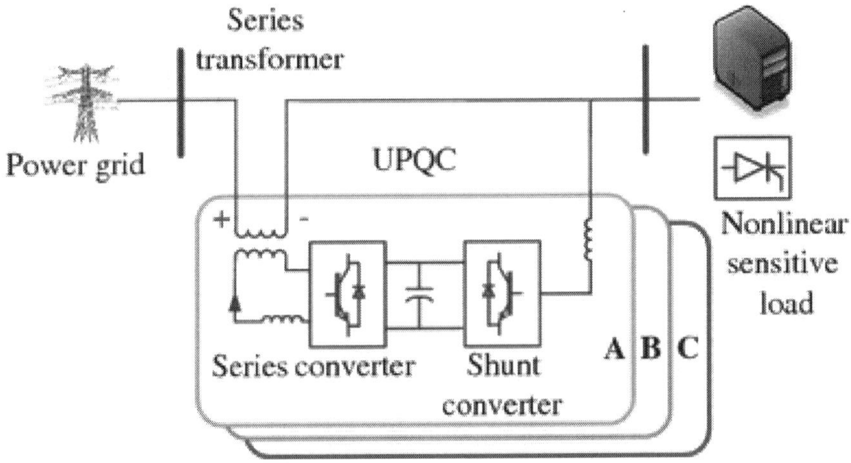

Figure 1. Structure of Unified Power Quality Conditioner

Voltage sag compensation strategy of UPQC can be divided into three types, namely pre-sag compensation, in-phase compensation, and minimum energy compensation [8–11]. The pre-sag compensation and in-phase compensation are similar to the compensation strategy of dynamic voltage restorer (DVR), which focus on minimum voltage magnitude and phase angle jump. Compare to minimum energy compensation of DVR, Ref. [12, 13] proposed a minimum VA method of UPQC (UPQC-VA_{min}), which utilizes non-linear optimum algorithm to calculate the injection angle of the voltage compensation so as to minimize the capacity of sag compensation. In addition, Ref. [14] proposed a voltage sag/swell compensation strategy with simultaneous active and reactive power injection by both converters of UPQC (UPQC-S), which can maximize the capacity utilization of series converter and compensate the load reactive power.

Both UPQC-VA_{min} and UPQC-S are based on the same principle, which determines a specific injection angle of injection voltage, for the sake of optimizing the output power of UPQC. Although these strategies can reduce the installed capacity of UPQC, they need to inject a large amount of active power, which will present difficulty in achieving sustainable compensation for voltage sag [15, 16]. However, for low voltage distribution system, on the premise of maintaining the voltage of the DC bus, UPQC usually has sufficient capability of voltage injection for voltage sag compensation. Therefore, instead of the capacity utilization, the current-carrying utilization becomes the key point of the optimized compensation strategy.

This paper proposed a simultaneous reactive power injection (UPQC-SRI) strategy for voltage sag compensation, which is based on the optimization of current-carrying utilization. This strategy uses both the series and shunt converters to inject reactive power for compensating voltage sag, and calculates the voltage injecting angle based on the minimization of current-carrying. What's more, the zero active power injection region of UPQC-SRI (also called UPQC-SRI region) and the current limit value for the case of exceeding the UPQC-SRI region are obtained. Meanwhile, the load harmonic mitigation is also considered, and the comparison between proposed strategy and UPQC-S is also performed.

2. PRINCIPLE OF UPQC-SRI VOLTAGE SAG COMPENSATION

The compensation phase diagram of UPQC-SRI when voltage sag occurs is shown in Fig. 2. With inductive load, for example, the series converter injects the voltage whose phase is ahead of the system. In the meantime, shunt converter injects reactive power to the system, which makes the voltage injection of series converter perpendicular to the system current. Therefore, UPQC can minimize the compensation energy and simultaneously inject reactive power by two converters. The voltage injection angle is represented as α. According to the law of cosine, the relationship of voltage vectors in Fig. 2 can be given by

$$V_{\mathrm{Se}}^2 + V_{\mathrm{S}}^2 + 2\cos\alpha V_{\mathrm{Se}} V_{\mathrm{S}} = V_{\mathrm{L}}^2 \tag{1}$$

where V_{Se}, V_S and V_L are the magnitude of compensation voltage, system voltage and load voltage, respectively. The compensation voltage can be derived as

$$V_{Se} = V_L(\sqrt{1 - D^2 \sin^2 \alpha} - D\cos\alpha) \quad (2)$$

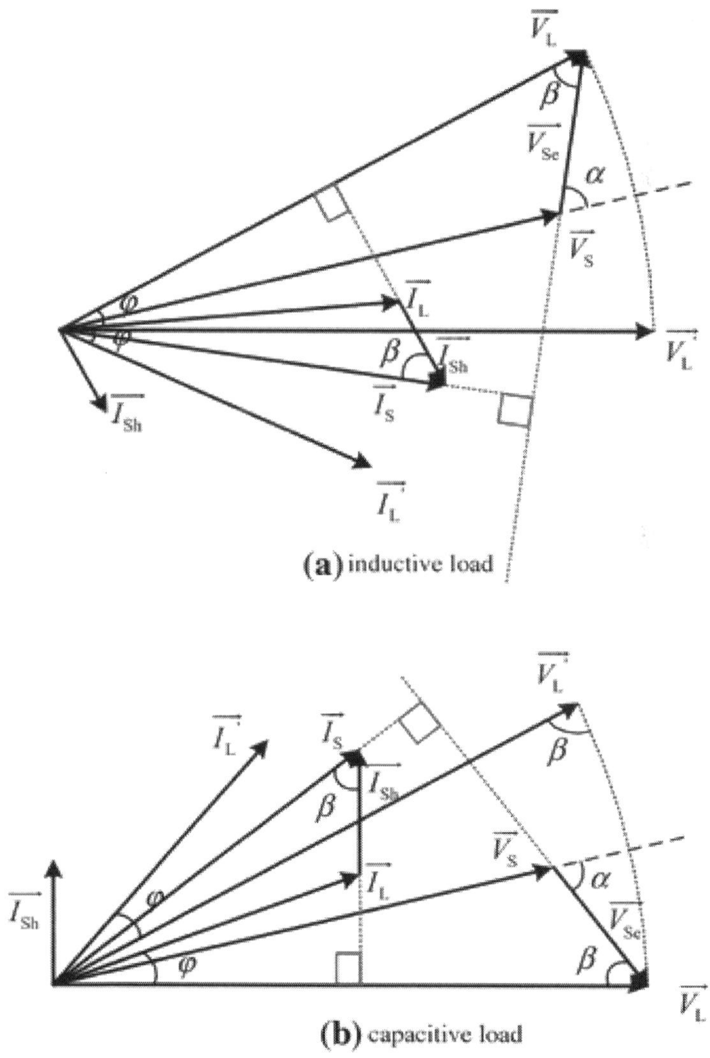

Figure 2. Vectogram of voltage sag compensation using UPQC-SRI

where D is the coefficient of voltage sag, and $D = V_s / V_L$. Meanwhile, the shunt converter injects reactive power for compensating voltage sag, and the magnitude of compensation current I_{Sh} can be given by

$$\begin{aligned} I_{\text{Sh}} &= I_{\text{L}} |\cot \beta \cos \varphi - \sin \varphi| \\ &= I_{\text{L}} \left| \frac{\sqrt{1 - D^2 \sin^2 \alpha}}{D \sin \alpha} \cos \varphi - \sin \varphi \right| \end{aligned} \tag{3}$$

where I_L represents the magnitude of load current, and φ is the power angle of load. The magnitude of system current I_S is given by

$$I_{\text{S}} = \frac{I_{\text{L}} \cos \varphi}{D \sin \alpha} \tag{4}$$

Then, the current-carrying of series converter I_{Se} is

$$I_{\text{Se}} = nI_{\text{S}} \tag{5}$$

where n is voltage ratio of series transformer of the UPQC.

As can be seen from (2)–(5), V_{Se}, I_S, and I_{Sh} are the functions of voltage injection angleα, of which V_{Se} increases with the increase of α, and I_S decreases when α increases. According to (3), I_{Sh} can be divided into two parts. The first half will increase with the increase of α, while the second part can be seen as constant. When

$$\frac{\sqrt{1 - D^2}}{D} \cos \varphi \le \sin \varphi \tag{6}$$

i.e. $D < \cos\varphi$, I_{Sh} increases with the increase of α. When $\alpha = 90°$, the current flowing through the series and shunt converters reaches to minimum value at the same time. Then, V_{Se} can be represented as

$$V_{\text{Se}} = V_{\text{L}}(\sqrt{1 - D^2}) \tag{7}$$

The current of shunt converter I_{Sh} is given by

$$I_{Sh} = I_L \left(\frac{\sqrt{1 - D^2}}{D} \cos \varphi - \sin \varphi \right) \tag{8}$$

The current which flows through series converter is

$$I_{Se} = \frac{n I_L \cos \varphi}{D} \tag{9}$$

On the contrary, when $D > \cos\varphi$, the current magnitude of the shunt converter can be zero by adjusting voltage injection angle α of the series converter. In this case,

$$\alpha = \arcsin\left(\frac{\cos \varphi}{D}\right) \tag{10}$$

This conclusion is consistent with the minimum energy compensation in [9].

The above conclusions are also suitable for the case of capacitive load. Under the consideration of different circumstances of voltage sag, α can be determined as 90° using the compensation strategy of USQC-SRI. The phase diagram is shown in Fig. 3. This method can realize zero active power injection of voltage sag compensation with minimized converter current-carrying, and can compensate the load reactive power. However, the aforementioned conclusions are deduced under an ideal condition without considering current limitation. So UPQC-SRI strategy with the constraint of current-carrying limit will be discussed in the following sections.

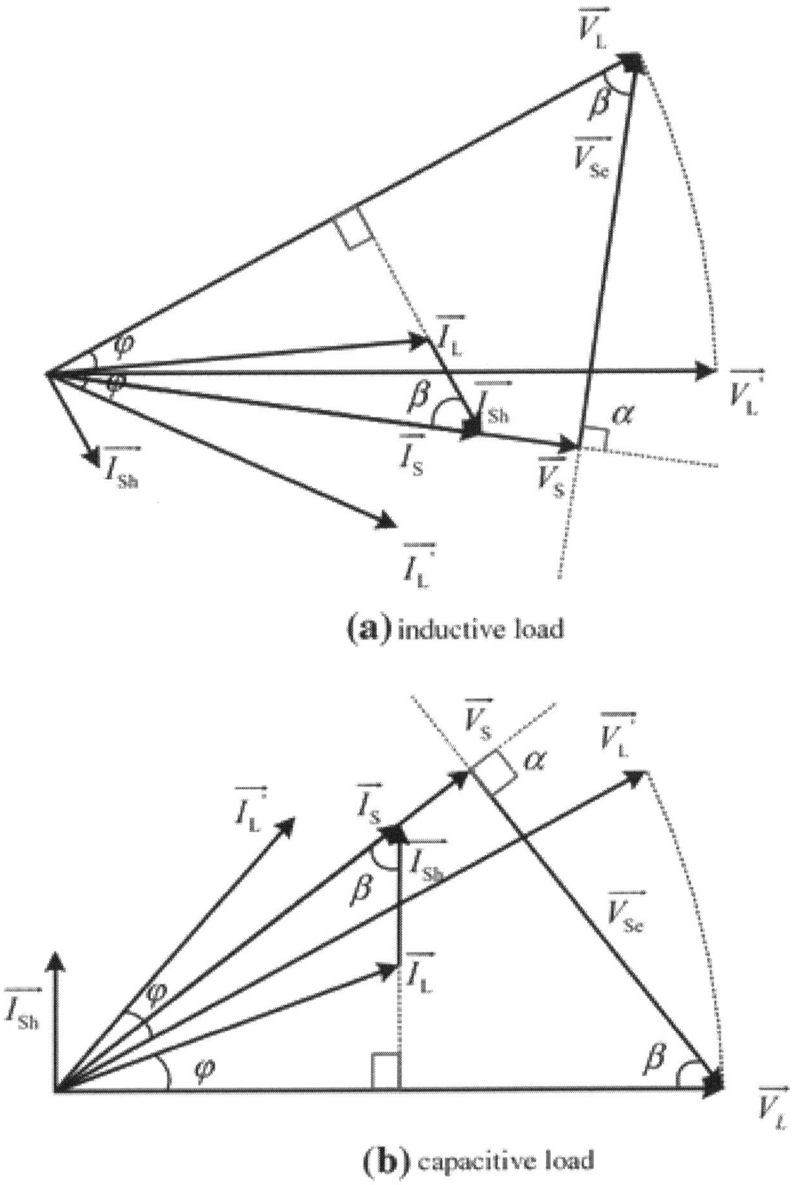

Figure 3. Vectogram of voltage sag compensation using minimum current method with zero active power injection

3. ANALYSIS OF ZERO ACTIVE POWER INJECTION REGION USING UPQC-SRI

3.1 Analysis of current-carrying requirements

Several researches mainly focus on capacity optimization of the UPQC [13–16]. However, the proposed strategy focuses on the limitation of current-carrying capability due to the following reasons.

1) Generally, UPQCs are directly installed at the non-linear and sensitive load at low voltage level. Taking 380 V three-phase system as an example, UPQC consists of three groups of single-phase full-bridge back to back converters using IGBTs as switches, as shown in Fig. 1. The DC bus voltage could exceed 600 V.

2) Using the proposed strategy, UPQC could maintain the voltage of dc bus without the output of active power (assuming the loss of UPQC could be ignored). Therefore, the series converter can operate in any circumstances of voltage sag without the limitation of the compensation voltage by properly designing the ratio of the series transformer.

3) The system current is the vectorial sum of the load current and the shunt converter compensation current. When severe voltage sag occurs, the magnitude of the system current will be larger than the load current, which could lead to the overcurrent of converters using UPQC-SRI. Based on (8) and (9), the current-carrying capability of UPQC has close relationship with the condition of the voltage sag and the voltage ratio of the series transformer.

As a result, current-carrying capabilities of both converters become the main factors of the voltage compensation. Specially, in section 3.1 to 3.3, the load harmonic mitigation is ignored, which will be elaborated in section 3.4. The current-carrying curves can be plotted as Figs. 4 and 5 based on (8) and (9).

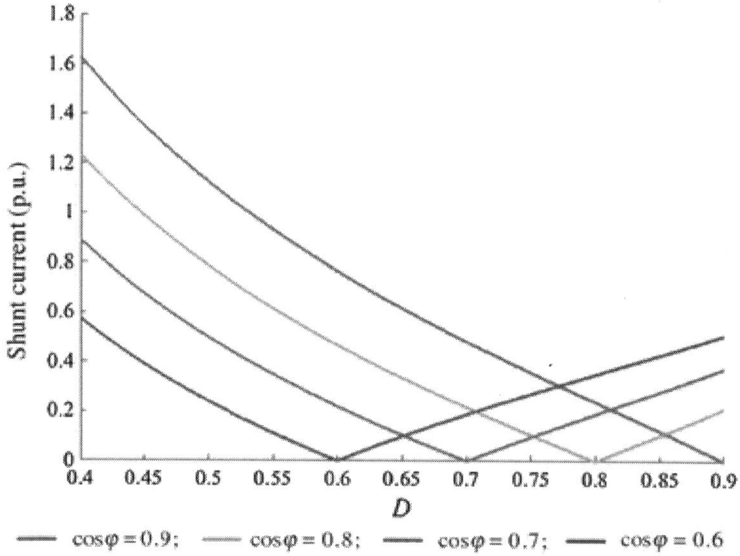

Figure 4. Curves of I_{sh} using UPQC-SRI under different voltage sag conditions

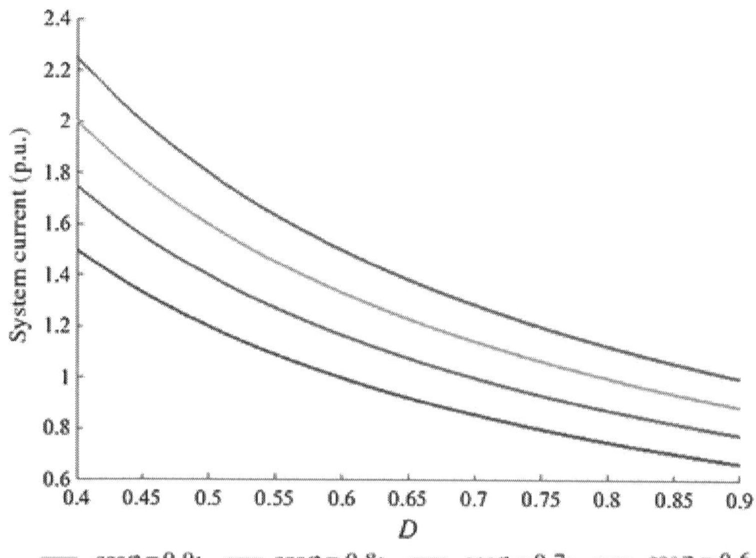

Figure 5. Curves of I_s using UPQC-SRI under different voltage sag conditions

Assume I_L equals to 1 p.u. and the ratio of the series transformer is 1:1. As shown in Figs. 4and 5, when $D = \cos\varphi$, I_{sh} achieves zero, and I_S equals to I_L. When $D > \cos\varphi$, I_S will be smaller than I_L because reactive current is injected from shunt converter with the opposite phase to the reactive component of load current during the voltage compensation using UPQC-SRI. When $D < \cos\varphi$, although UPQC-SRI could compensate the voltage sag by reactive power injection, the shunt converter will inject reactive current with the same phase as the reactive component of load current. Then, the magnitude of system current will become larger than the load current. As the result, I_S is likely greater than I_L using UPQC-SRI, which would cause the overcurrent of UPQC.

With consideration of the cost of UPQC, the two converters have the same insulation level, current-carrying ability and capacity. In consequence, it is necessary to properly design the transformer voltage ratio to enlarge the UPQC-SRI region.

3.2 UPQC-SRI region and current limit

The voltage ratio of the UPQC series transformer has close relationship with UPQC-SRI region. If there is no limitation of the voltage injection magnitude, then the constraint of the series coupled transformer voltage ratio is

$$n < \frac{V_L\sqrt{1-D^2}}{m_c V_{DC}} \tag{11}$$

where m_c is the modulation ratio and V_{DC} is the DC bus voltage of UPQC. Assume $V_{DC} = 600$ V with the consideration of cost and insulation, $D = 0$, $m_c = 0.85$, and $n < 0.61$ under the circumstance of severe voltage sag. The voltage ratio of the transformer can be set to 3:5, and the following discussion will be based on these settings of the parameters.

Let m_i denote the current-carrying ratio of series and shunt converter,

$$m_i = \frac{I_{Sh}}{I_{Se}} = \frac{\sqrt{1-D^2}\cos\varphi - D\sqrt{1-\cos^2\varphi}}{n\cos\varphi} \tag{12}$$

Assume that the current-carrying capacity of UPQC equals to the rated load current magnitude, and then UPQC-SRI region can be obtained as Fig. 6 based on (8), (9) and (12).

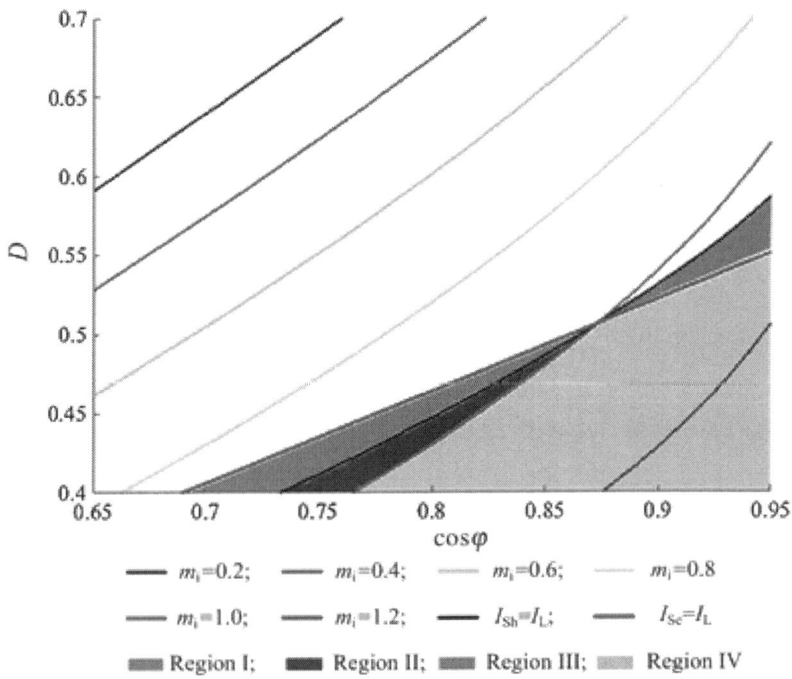

Figure. 6. Contour map of UPQC-SRI region

The cases when $I_{Se} = I_L = 1$ p.u. and $I^* = I_{Sh} = I_L = 1$ p.u. are illustrated in Fig. 6 in black and red lines, where I^* is the current capacity of UPQC. The contour curves of m_i at $n = 0.6$ are displayed. As can be known from Fig. 6, the area without shadow is UPQC-SRI region, which could compensate the voltage sag with simultaneous reactive power injection of both converters. When the voltage sag occurs in the shadow area, it will cause overcurrent using UPQC-SRI. According to the current magnitude of two converters and the overcurrent conditions, the shadow area could be divided into four regions(Region I, II, III and IV).

1) Region I: When the voltage sag happens in this area, the current-carrying of the series converter will be larger than the shunt converter, and the overcurrent will occur at the series converter. In order to avoid the overcurrent of the series converter, I_{Sh} needs to be limited, which can be represented by I_{Sh_limit}

$$I_{Sh\_limit} = \left| \sqrt{\left(\frac{I^*}{n}\right)^2 - (I_L \cos\varphi)^2} - I_L \sin\varphi \right| \tag{13}$$

2) Region II: When the voltage sag occurs in this area, both the series and shunt converters would exceed their current limits using UPQC-SRI, and then the current of shunt converter needs to be clamped. The amplitude of current limit is I_{Sh_limit}.

3) Region III: In this area, the current-carrying of series converter will be smaller than the shunt converter. Only the current of shunt converter will exceed its limit. Then the current limitation of shunt converter should be I^*.

4) Region IV: In this area, the current-carrying of series converter will be smaller than the shunt converter. Both converters will exceed their current limit, and I^* should also be the limit of shunt converter.

As the result, compared with the conclusion (namely $D > \cos\varphi$) of [9], UPQC-SRI can obtain a larger region of zero active power injection with proper design of n.

3.3 Comparative analysis of active power injection

The clamping current of shunt converter is required using UPQC-SRI strategy. When the shunt current reaches its limit, the phase diagram of voltage compensation is shown in Fig. 7. Due to the current limit, the system current will be no longer perpendicular to the injected voltage of series converter, and the angle is γ. It implies that series converter injects active power P_{inj} to the load, where

$$\begin{aligned} P_{\text{inj}} &= \overrightarrow{V_{\text{Se}}} \cdot \overrightarrow{I_{\text{S}}} = V_{\text{Se}} I_{\text{S}} \cos\gamma \\ &= V_{\text{L}} \sqrt{1 - D^2} (I_{\text{L}} \cos\varphi - D I_{\text{Sh\_limit}}) \end{aligned} \quad (14)$$

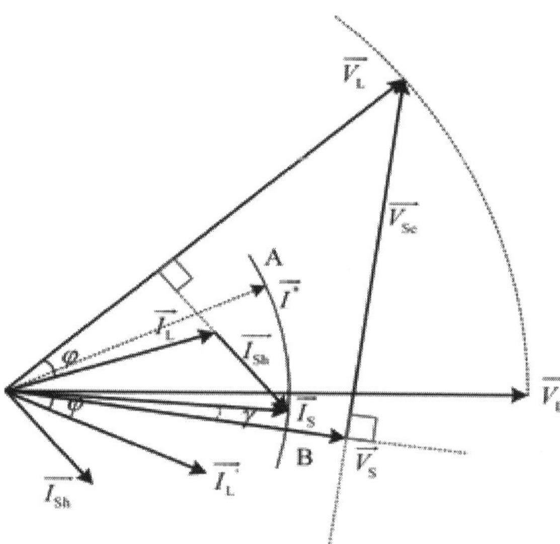

Figure 7. Vectogram of voltage sag compensation simultaneously considering the current-carrying limit of UPQC

According to (14), the more severe voltage sag occurs, the more active power is required from UPQC.

Compared with [14], the demand of active power injection using UPQC-S strategy P_{inj}` can be given by

$$P'_{inj} = V_L I_L \left[\cos(\varphi - \delta_{max}) + \frac{\cos \delta_{max}}{D}\right] \quad (15)$$

where δ_{max} is the maximum difference value of phase angle between the pre-compensated and compensated load voltage. Assuming the voltage limit of V_{Se} for UPQC-S and UPQC-SRI is the same, then $\cos\delta_{max} = 0.5$ can be obtained. When the voltage sag occurs in region III or region IV, the active power injection ratio of UPQC-SRI to UPQC-S is defined as m_p, which can be given by

$$m_p = \frac{P_{inj}}{P'_{inj}} = \frac{\sqrt{1 - D^2}(\cos \varphi - D)}{\cos(\varphi - \delta_{max}) + \frac{\cos \delta_{max}}{D}} \quad (16)$$

According to (16), the ratio of output active power is shown in Fig. 8 with contour curve ofm_p. When the sag happens in the shadow area, zero active power will be injected for the voltage sag compensation using UPQC-SRI. When the severe sag occurs, the demand for active power injection using UPQC-SRI is also smaller than that one using UPQC-S. Therefore, UPQC-SRI can provide more sustaining compensation for the voltage sag.

3.4 Consideration of harmonic mitigation

As the comprehensive compensation equipment, UPQC is able to mitigate both current and voltage power quality issues. However, there may be a conflict when mitigating voltage sag and load harmonic current at same time using UPQC-SRI, which will narrow down the UPQC-SRI region or cause the overcurrent of shunt converter. Therefore, it becomes much more complicated to deal with these two power quality issues at same time. Generally, the comprehensive mitigation of voltage sag and load harmonic current could be divided into three conditions.

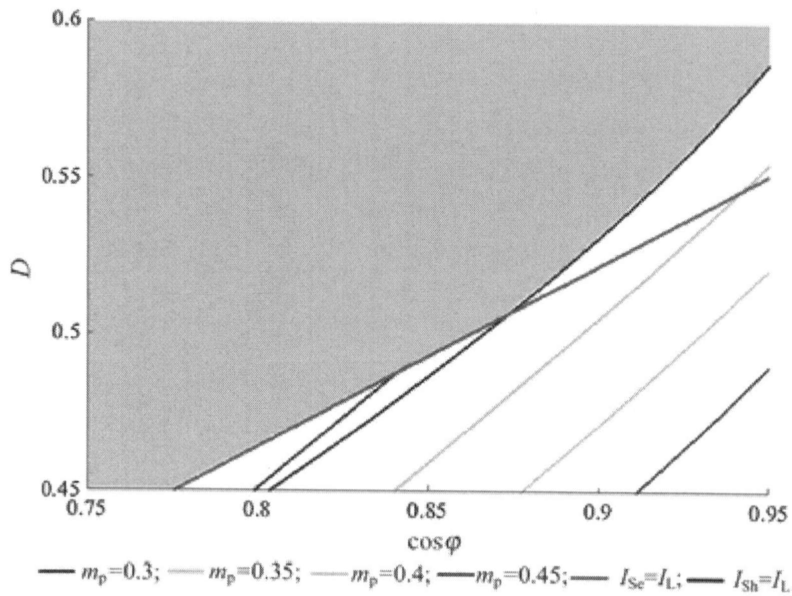

Figure 8. Active power injection comparison of UPQC-S and UPQC-SRI during voltage sag

1) Condition I: When the voltage sag occurs in UPQC-SRI region, the mitigation of voltage sag and load harmonic current can be performed at the same time. The overcurrent will only happen in shunt converter if the harmonic component of load current is very large. Therefore, the current limit of shunt converter can be represented as *I**.

2) Condition II: When the voltage sag occurs in region I or II, the current-carrying of series converter will become larger than that of shunt converter. Thus, the shunt converter may have extra current capacity for the mitigation of load harmonic current using UPQC-SRI. Under this condition, the current injection of shunt converter should be limited.

Different from Condition I, the fundamental component of shunt current should also be clamped at a proper value, which can avoid the overcurrent of both series and shunt converters. It can be given by

$$\begin{cases} I_{\mathrm{Sh\_limit\_f}} = \left| \sqrt{\left(\frac{I^*}{n}\right)^2 - (I_{\mathrm{L}} \cos\varphi)^2} - I_{\mathrm{L}} \sin\varphi \right| \\ I_{\mathrm{Sh\_limit}} = I^* \end{cases} \quad (17)$$

where $I_{\mathrm{Sh\_limit\_f}}$ is the clamped value of fundamental component of shunt current.

3) Condition III: When the voltage sag condition occurs in region III or IV, the current-carrying of series converter will be smaller than that of shunt converter, and the current limit of the shunt converter is represented as I^*. Thus, the shunt converter may not have extra current capacity for mitigation of load harmonic current. Therefore, under this condition, the mitigation of load harmonic current should turn off during the voltage sag period.

4. UPQC-SRI ALGORITHM AND ITS VALIDATION

4.1 Compensation algorithm

The block diagram of the closed-loop control algorithm is shown in Fig. 9.

For the closed-loop control of series converter, the vector of system voltage is constructed in αβ coordinate using delay operation [17]. Then the system voltage can be transformed into dq coordinate. After that, the system voltage will be transformed to DC component on d axis. Then, *D* can be calculated in real-time property [18]. Since the compensation voltage is perpendicular to the system voltage, the transformed compensation voltage only exists in q axis. Based on above calculation, the modulation voltage is obtained, which needs to be transformed to the abc coordinate. The difference between referenced compensation voltage and output compensation voltage caused by series transformer needs to be adjusted using proportion resonance (PR) controller. The resonant frequency of PR controller is set at 50 Hz in this paper [19].

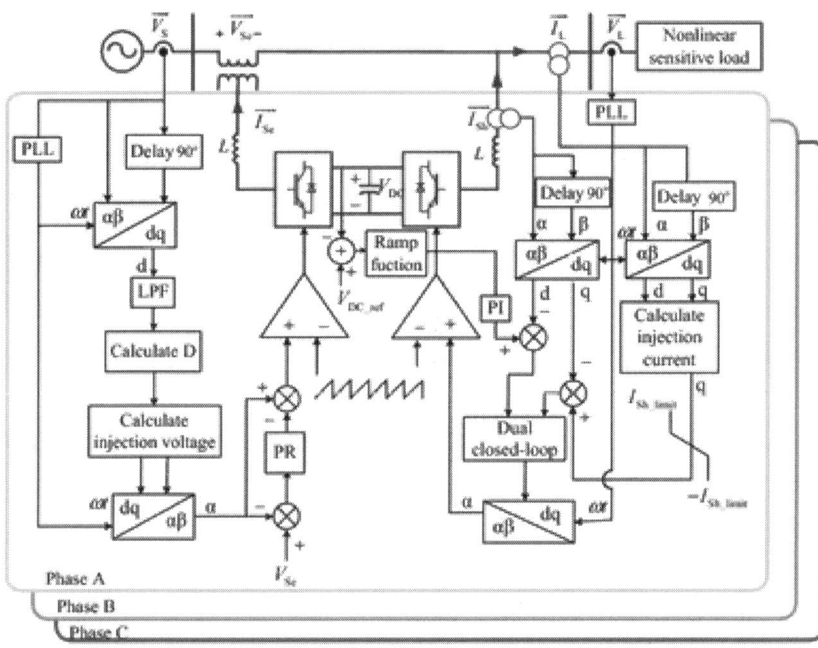

Figure 9. Per-phase control block diagram of voltage sag simultaneously compensation

For the shunt converter, the same transformation is used to calculate the shunt current. The elements of $I_L\cos\varphi$ and $I_L\sin\varphi$ in (8) are transformed to the DC components in d and q axis, respectively. As a result, load power factor could be measured in real-time property. Therefore, the detection delay of load switching can be determined by the design of digital filter. Furthermore, the pulse width modulation can be generated by classical dual closed-loop control. In order to achieve the stationary charge-discharge of DC capacitance, a ramp function is used in DC bus voltage control. Specially, a saturate block in the q axis is demanded due to the current-carrying limit. DC voltage control should be stopped during voltage sag period. And then the UPQC will not absorb active power, which may cause more severe voltage sag.

4.2 Simulation results

In order to verify the validity of UPQC-SRI, a PSCAD/EMTDC simulation model is constructed. The parameters of simulation are shown in Table 1.

In this simulation, voltage sag occurs at 1.5 s with D = 0.6, and the duration is 150 ms. The current limit of the UPQC is set at 18 A (peak). The proposed strategy is tested under two load conditions, which are in and out of the UPQC SRI region, respectively.

4.2.1 Condition 1

The root mean square (RMS) value of load phase current is 14 A, and the power factor is 0.74. The theoretic output current of the shunt converter is 4.4 A, which is smaller than the limitation 9.07 A according to (8) and (12). Therefore, this voltage sag occurs in the UPQC-SRI region, and the waveform is shown in Fig. 10.

As shown in Fig. 10, before the voltage sag happens, injection voltage of series converter equals to zero and the shunt converter provides reactive power compensation for the system. The voltage recovers to the normal rating within half cycle after voltage sag occurs. The angle of voltage injection of the series converter is 90 degrees ahead of system voltage due to the inductive load. Meanwhile, the shunt converter injects inductive reactive power so as to make the system current perpendicular to injection voltage. Then, the system voltage is in phase with system current. Because there is no active power exchange, the DC bus voltage of UPQC drops a little. After sag period, the DC voltage returns to the rating under the control of DC voltage.

Table 1. Parameters of the simulation systems

Parameter	Value
Rate system voltage (RMS) V_S	380 V
UPQC DC voltage V_{DC}	600 V
DC capacitance C	4700 μF
inductance L	6.3 mH
Series transformer capacity	40 kVA
Series transformer leakage reactance	0.08 p.u.
Series transformer voltage ratio n	3:5
PR controller resonant frequency	50 Hz
PR controller proportional factor	80
PR controller resonant factor	0.01
Current inner loop proportional factor	20
Current inner loop integral factor	0.005
dq digital filter type	Butterworth
dq digital filter order	2
dq digital filter cut-off frequency	50 Hz

4.2.2 Condition 2

In this condition, load phase current is 16 A (RMS) and power factor is 0.96. According to (8), the theoretic minimum value of zero active power injection current of the shunt converter is about 16 A (RMS), which will lead to the overload of the series converter.

This condition of voltage sag exceeds the UPQC-SRI region, and the waveform is shown in Fig. 11. According to Fig. 6, the voltage sag condition occurs in region I, in which the series converter overload may happen. According to (13), the limit current of shunt converter is 10.15 A (peak). Due to the active power injection, the DC bus voltage no longer sustains. After the voltage sag period, the dc bus voltage returns to normal rating gradually.

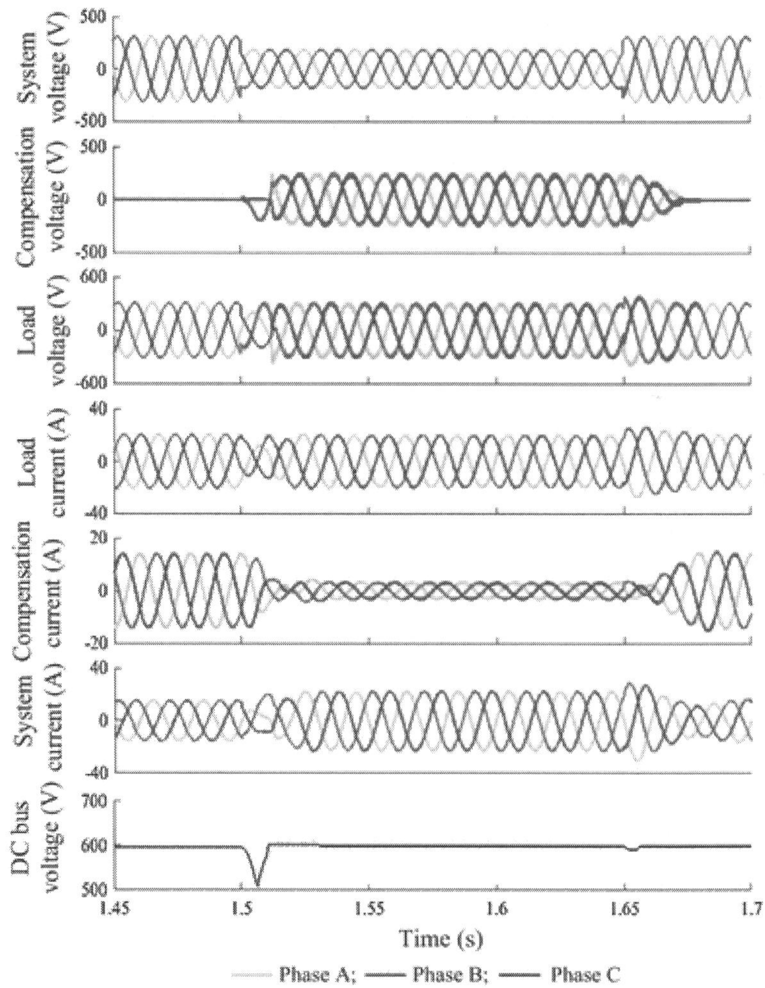

Figure 10. Simulation results of UPQC-SRI under condition 1

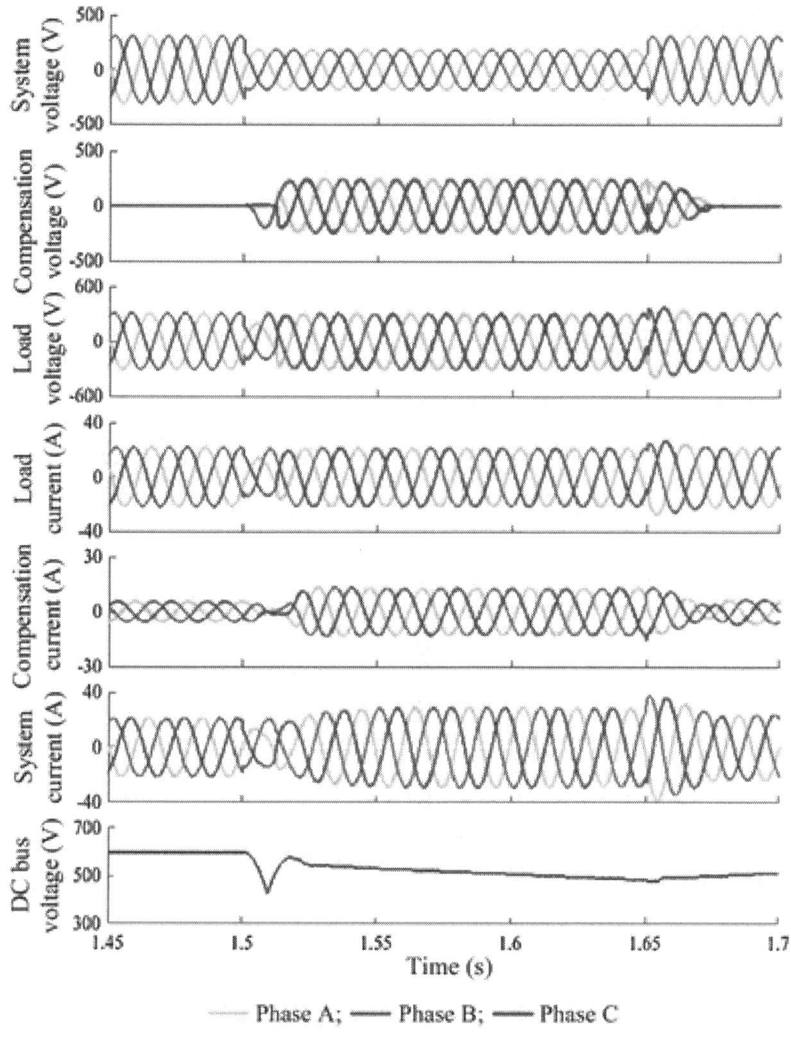

Figure 11. Simulation results of UPQC-SRI under condition 2

4.2.3 Transient response and algorithm comparison

In order to ensure the real-time property of corresponding control algorithm, the PR controller is utilized to track the sinusoidal reference voltage. Meanwhile, the real-time sampling of power factor is performed. The simulation result is represented in Fig. 12.

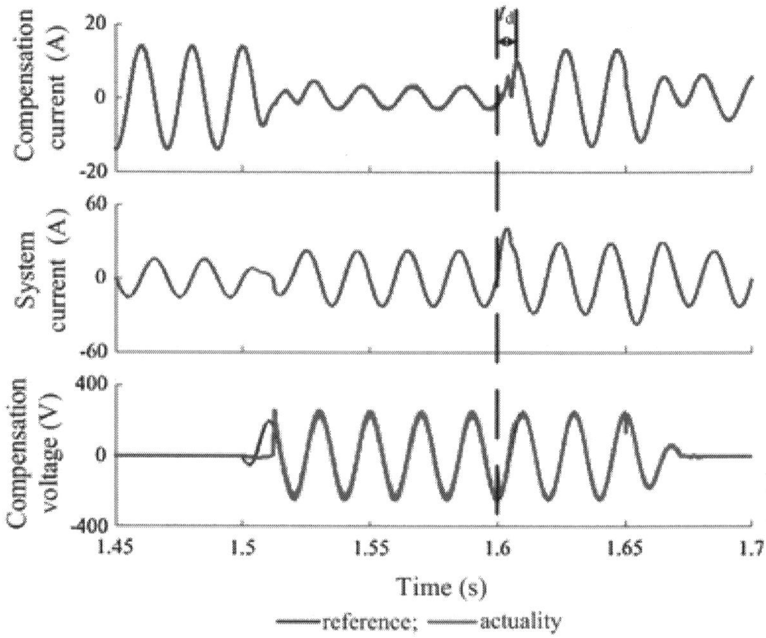

Figure 12. Dynamic simulation results of UPQC-SRI

The simulation starts from condition 1, and the transient of the load switching occurs at 1.6 s. Then, the load condition switches to condition 2. The result indicates that the PR controller can accurately and quickly track the reference voltage, and the transient response of shunt current t_d is shorter than 10 ms.

As can be seen from Fig. 13, the voltage of DC bus drops more deeply using UPQC-S, which means the more active power injection than UPQC-SRI during voltage sag. Therefore, the analysis in section 3.3 can be verified by the simulation results. Compared with UPQC-S, the proposed strategy can provide more sustainable voltage sag compensation.

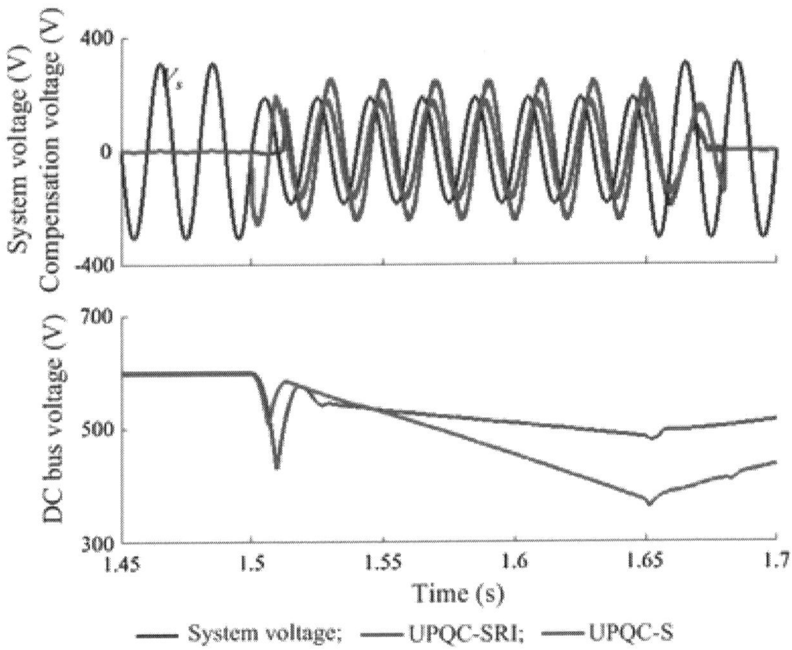

Figure 13. Simulation comparison between UPQC-S and UPQC-SRI

5. CONCLUSION

Firstly, the principle of voltage sag compensation is derived in this paper. And then UPQC-SRI voltage compensation strategy with optimal configuration of current-carrying is proposed. This strategy can achieve the control of unity power factor and provide sustainable voltage sag compensation by simultaneous reactive power injection from both shunt and series converters of the UPQC.

With consideration of current-carrying limit, the UPQC-SRI region is presented, and then the limit value of the shunt current is obtained. Furthermore, the conditions for compensating both the voltage sag and the load harmonic current are analyzed under three sag conditions.

Finally, the UPQC-SRI strategy and the corresponding algorithm are verified by PSCAD/EMTDC platform. Both the analysis and the simulation results indicate that when the severe voltage sag occurs, UPQC-SRI can compensate the voltage sag with less active power injection than using the traditional strategies.

ACKNOWLEDGEMENTS

This work was supported by the twelfth five-year National Mega-projects of Science and Technology (2011BAA01B03).

REFERENCES

1. Xiao XN (2004) Analysis and control of power quality. China Electric Power Press, Beijing, pp 124–132 (in Chinese)

2. Guo XQ, Wang HB, Lu ZG et al (2014) New inverter topology for ground current suppression in transformerless photovoltaic system application. J Mod Power Syst Clean Energ 2(2):191–194

3. Zhong Q, Gao XH, Yu NH et al (2014) Accommodating capacity and mode of distributed generation under harmonic constraint in active distribution networks. Autom Electr Power Syst 38(24):108–113

4. Fujita H, Akagi H (1998) The unified power quality conditioner: the integration of series and shunt-active filters. IEEE Trans Power Electron 13(2):315–322

5. Ganguly S (2014) Impact of unified power-quality conditioner allocation on line loading, losses, and voltage stability of radial distribution systems. IEEE Trans Power Deliv 29(4):1859–1867

6. Jothibasu S, Mishra MK (2014) A control scheme for storageless DVR based on characterization of voltage sags. IEEE Trans Power Deliv 29(5):2261–2269

7. Zhang ZH, Xu BY, Chen Q (2012) Control strategies for UPFC-based optimal power flow of distribution network with normally closed-loop operation. Power Syst Technol 36(6):122–126 **(in Chinese)**

8. Long YB, Xu YF, Xiao XN et al (2015) Pre-charging control for unified power quality conditioner based on modular multilevel converter. Autom Electr Power Syst 39(7):182–187

9. Liu YY, Xiao XN, Xu YH (2010) Characteristics analysis on energy steady compensation for dynamic voltage restorer. Proc CSEE 30(13):69–74 (in Chinese)

10. Wang SW, Huang K, Xu XT (2014) Control strategy for dynamic voltage restorers without inner current loops. Autom Electr Power Syst 38(18):93–98

11. Khadkikar V, Chandr A, Barry AO et al (2011) Power quality enhancement utilizing single-phase unified power quality conditioner: digital signal processor-based experimental validation. IET Power Electron 4(3):323–331

12. Rauf AM, Khadkikar V (2015) An enhanced voltage sag compensation scheme for dynamic voltage restorer. IEEE Trans Ind Electron 62(5):2683–2692

13. Kumar GS, Vardhana PH, Kumar BK et al (2009) Minimization of VA loading of unified power quality conditioner (UPQC). In: Proceedings of the international conference on power engineering, energy and electrical drives (POWERENG '09), Lisbon, Portugal, 18–20 March 2009, pp 552–557

14. Khadkikar V, Chandra A (2011) UPQC-S: a novel concept of simultaneous voltage sag/swell and load reactive power

compensations utilizing series inverter of UPQC. IEEE Trans Power Electron 26(9):2414–2425

15. Bhavani R, Prabha NR, Kanmani C (2015) Fuzzy controlled UPQC for power quality enhancement in a DFIG based grid connected wind power system. In: Proceedings of the 2015 international conference on circuit, power and computing technologies (ICCPCT '15), Nagercoil, India, 19–20 March 2015, 7 pp

16. Ganguly S (2014) Impact of unified power-quality conditioner allocation on line loading, losses, and voltage stability of radial distribution systems. IEEE Trans Power Deliv 29(4):1859–1867

17. Xiao XN, Xu YH, Liu H (2002) Research on the detection method of voltage sag characteristics. Electr Power Autom Equip 22(1):19–22 (in Chinese)

18. Xiao XN, Xu YH, Liu LG (2002) Research on mitigation methods of voltage sag with phase-angle jump. Proc CSEE 22(1):64–69 (in Chinese)

19. Xu SH, Li JL, Hui D (2015) Stability analysis of energy storage inverter based on quasi PR controller under off-grid mode. Autom Electr Power Syst 39(19):107–112

Chapter 8

Hybrid wind power balance control strategy using thermal power, hydro power and flow batteries

Linas Gelazˇanskas [a], Audrius Baranauskas [b,c], Kelum A.A. Gamage [a]. Mindaugas Azˇubalis [b]

[a]Department of Engineering, Lancaster University, Bailrigg, Lancaster LA1 4YR, UK

[b]Department of Power Systems, Kaunas University of Technology, Studentu Str. 48, Kaunas LT-51367, Lithuania

[c] Strategy and Research Division, LITGRID AB, A. Juozapaviciaus Str. 13, Vilnius LT-09311, Lithuania

ABSTRACT

The increased number of renewable power plants pose threat to power system balance. Their intermittent nature makes it very difficult to predict power output, thus either additional reserve power plants or new storage and control technologies are required. Traditional spinning reserve cannot fully compensate sudden changes in renewable energy power generation. Using new storage technologies such as flow batteries, it is feasible to balance the variations in power and voltage within very short period of time. This paper summarises the controlled use of hybrid flow battery, thermal and hydro power plant system, to support wind power plants to reach near perfect balance, i.e. make the total power output as close as possible to the predicted value. It also investigates the possibility of such technology to take part in the balance of the Lithuanian power system. A dynamic model of flow battery is demonstrated where it evaluates the main parameters such as power, energy, reaction time and efficiency. The required battery size is tested based on range of thermal and hydro power plant reaction times. This work suggests that power and energy of a reasonable size flow battery is sufficient to correct the load and wind power imbalance.

KEYWORDS

Flow battery; Power system balance; Wind farms; Wind power generation

1. INTRODUCTION

During the past decade, the number of renewable energy sources has increased dramatically. It is forecasted that the growth of green energy generation will increase even further. Policy makers in developed countries create many incentives in favour of the development of low-carbon technologies and subsidise green energy generation. This should help to reduce carbon footprint and climate change. On the other hand, most of renewable energy comes from generators that are inherently very hard to control [1], thus it introduces further complexity in system balancing task.

Up to now in many cases wind turbines or solar panels are being connected to the grid with minimal control. Due to hardly predictable natural resources, like wind or solar irradiation, the errors between actual energy output and forecasted generation are relatively large. This increases the difficulty of the energy balance problem: corresponding operators need either more tools or new technologies to come in hand [2]. Increasing advanced spinning reserve to back up intermittent generation would require inadequate level of investment considering exponential growth of power generation using green technologies. Also, this type of reserve has limited power variation capabilities (in the order of minutes) whereas solar power output can drop nearly instantly. The alternative is to use new highly responsive storage technologies [3] and [4]that could be incorporated into the system and shave over-generation as well as generate energy when demand overtakes supply.

Lithuanian Power System (PS) and other Baltic States currently operate synchronously with IPS/UPS synchronous zone and are connected to BRELL power ring, which consists of Belarus, Russian, Estonian, Latvian, and Lithuanian power systems. According to BRELL regulations Baltic States are not required to have automatic secondary power control, however Baltic States are planning to synchronically connect to the power grid of Continental Europe in 2020. This means decentralisation of power system control and responsibility to maintain power and energy balance within strict boundaries [5]. Therefore, it is important to investigate the feasibility of Lithuanian PS to automatically maintain power balance.

National Renewable Energy Laboratory in USA focuses on researching economic feasibility of energy storage and clearly states that high penetration of variable generation increases the need for all flexibility options including storage [6]. They also note that the economic value of energy storage devices is at its best when selling to the entire grid, instead of any single source. However the role of storage for wind generation requires continuous analysis and additional studies including new techniques to evaluate more dynamic grid operation.

Droste-Franke et al. analysed German power system balancing options and concluded that technological progress is needed in the following areas. Firstly, grid expansion and inter-regional connectivity compensating regional shortages of supply from renewable sources in Europe. Secondly, load management could become feasible through technologies such as smart metering, and finally, storage capacities need to be extended[7]. They also conclude that interrupted renewable power smoothing using battery storage system [4] and [8] is the cheapest option at present.

Söder and Hamon investigate wind power capabilities to participate in balancing services. They conclude that wind power plants do not usually participate in balancing services because they must be set to produce less than they are capable in order to be used for up regulation [9]. Margins are kept by spilling the wind, which cannot be stored. A method is proposed to select a certain tertiary reserve control in order to minimise the total cost of the system and maintain stability of the power system with larger portions of wind power. This means that they deal with emergency power system operation modes while our proposed method covers secondary control reserve and optimal share of reserve power between different kinds of generation sources.

Lubosny and Bialek proposed wind farm supervisory control scheme which is suitable to control individual wind mill or separate wind farm in two different ways – using additional storage device or power reserve achieved through part-loading one or more turbines in a wind farm. Authors suggest using wind power filter in order to separate the variability of wind power. They also concluded that elimination of larger power variations can be done more effectively using a central or single energy storage [10]. Therefore our proposed control strategy differs due to the fact that it deals with central control of all wind farms instead of individual ones.

Jiang and Wang similarly to [10] suggest to control wind power plant using power filter. Additionally, they proposed the optimisation model of corresponding filter parameters. However, due to the uncertainty inherent in wind power generation, optimal control during long time periods has difficulties predicting wind power and is unpractical in actual real-time operation [11]. Besides it requires additional computational resources and time. Active power losses and state of charge of storage devices depend on wind power generation, therefore it is hard to maintain the proper charge level and mitigate wind power fluctuation. Authors conclude that two-time-scale coordination control method gives controversial results because the required battery power reaches 33% of installed wind power (in our case it reaches 5–25% depending on power system operation mode, discussed later in the paper) while the power fluctuation allowance is up to 10%. Finally, the capacity component of the battery dominates (comparing to power) which means that storage devices are controlled according to wind power trends. The control strategy proposed in this paper controls storage devices according to high frequency component of wind power imbalance and it allows reaching 100% power balance with reasonably lower ratings of storage device.

Abbey et al. suggests using filters and neural networks to control two different types or multiple storage devises [12]. It is novel and interesting method but too complex for wind power balancing purposes on real-time operation. In principle, multiple levels of storage is needed only in new areas such as island household networks with renewable sources or micro-grids while wind power integration to conventional power systems usually require only short-term storage because the trends of wind power imbalance could be compensated by thermal or hydro power plants in more economical way. Our proposed hybrid wind power balance control strategy composes of conventional generation and energy storage control from power system operator point of view, which means central control in more efficient manner.

Wang et al. studied operational reliability of power system with high wind power penetration [13]. They have concluded that energy storage systems dramatically increase reliability of systems with wind farms. Authors also analyse and show the required battery sizing for certain reliability index.

Khalid and Savkin proposed new semi-distributed storage configuration [14] and using model predictive control [15] identified the optimal capacity of battery energy storage system. However the purpose was to identify the optimal capacity only taking into account the system ramp rates while our paper also deals with installed power of energy storage, conventional power plants optimal control and active power reserve optimisation. In addition we have used actual wind data of 10 days with a time step of one second instead of 1 day and 10 min time step. Yuan et al. proposed dual-battery energy storage system [16] which consist of two separate battery storage systems. One of them is suitable for positive error compensation where the other one is suitable for negative ones. They also proposed three indices for the assessment of the performance on wind power dispatchability which could be identified by using sequential Monte Carlo simulation. However the time step is one hour which means that little dynamic behaviour could be represented. The authors also do not introduce any optimisation task.

Ansari and Velusami have been investigating the dynamic stability of hybrid autonomous wind – diesel with battery energy storage system. They proposed dual mode linguistic hedge fuzzy logic controller [17] and have shown its advantages comparing to traditional fuzzy logic and PI controllers. Kalantar and Mousavi G. replaces less effective and high pollution diesel generator to more flexible and reliable microturbine with the addition of solar array system to earlier investigated one. In order to maximise power outputs of wind and solar power plants they proposed a model reference adaptive Lyapunov controller[18] and improve the system behaviour comparing to fuzzy logic and PID controllers. Latter Mousavi G. have adapted the proposed method to offshore wind and tidal hybrid system with microturbine and BESS [19]. The authors provide an in depth investigation/review of the autonomous rural hybrid system in literature [17],[18] and [19], however the proposed methods are not suitable for wind integration to large power systems with conventional generation, therefore our paper deals with this issue. In addition, our paper represents the optimal wind balancing power allocation between conventional power plants and energy storage devices.

A feasibility study of hybrid solar-wind-battery system for remote location can be found in[20]. Although it shows that it is possible to replace diesel generators by 100% renewable energy, about 48.6% of energy is dumped due to lack of storage and energy management.

Traditional Automatic Generation Control (AGC) system calculates error of the control area and allocates the required regulating power plants. Then they participate in the system balance according to participation factors [21] in order to keep power system in balance. The participation factors are usually determined according to power plant's parameters such as rate limits [22] and [23], available spinning reserve [24] or economic (cost) characteristics. There are many methods to determine them: major part of power is allocated to the cheapest power plant, the fastest response power plant [25] or combined[21] method. This paper describes energy management method for increasing the quality of wind energy output using conventional Thermal Power Plant (TPP), Hydro Power Plant (HPP) and Flow Batteries (FB).

More proposed ideas of balancing wind power intermittency using energy storage systems can be found in other publications [26], [27] and [28]. In [3] authors show through simulation how flywheel ESS can be used for wind power smoothening.

2. THE MAIN FEATURES OF THERMAL AND HYDRO POWER PLANTS

In general the limits of power changing rate of TPP and HPP are quite different. The maximum load rate of TPP is about 2% of the installed unit capacity (per minute) while for HPP the maximum load rate could reach 100% of unit size (per minute). However it is impossible to perfectly match the area control error continuously in relation to variations in plants characteristics such as the system frequency, load or wind power plant output. The energy storage devices, such as flow batteries, could significantly improve the flexibility of the system control and reduce the power systems imbalances.

The following traditional transfer function of a classical hydraulic turbine could be expressed:

$$W_H(s) = \frac{\Delta\overline{P_m}}{\Delta G} = \frac{1 - T_w s}{1 + 0.5T_w s} \quad (1)$$

where $\Delta\overline{P_m}$ is the change in turbine mechanical power, ΔG is the change in gate position, and T_w is the water time constant.

Equation shows how the turbine output power reacts to a change in position of gate. Fig. 1 clearly shows that the initial power output of hydraulic turbine is two times opposite the value when the gate is opened immediately. This is due to water inertia which is represented by T_w. The complete response of hydraulic turbine in respect to gate step change can be seen in Fig. 1.

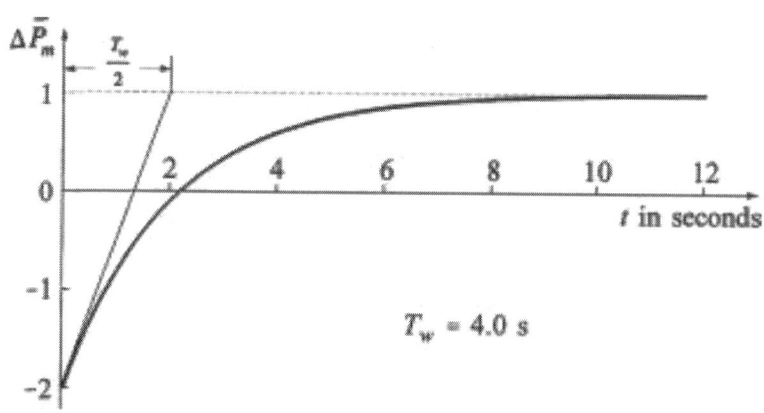

Figure 1. Hydraulic turbine mechanical power output response to gate position step change [29].

Fig. 2 is included in order to demonstrate the behaviour of main variables of HPP, when the gate position change is a ramp function during one second. It represents the main HPP parameters – head, power output and water velocity.

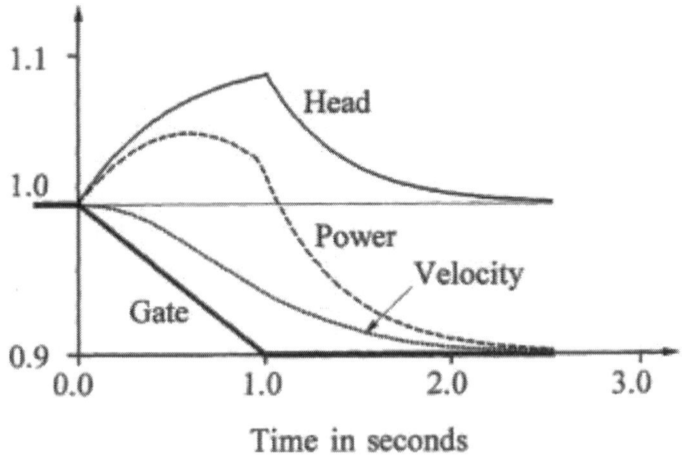

Figure 2. HPP main parameters respect to reduction in gate opening [29].

In order to get stable operation of HPP it is necessary to have permanent and temporary droop compensation when implementing the governors of the hydraulic turbine. The result is a governor with a high droop for fast speed deviations and a low droop for steady state [30].

The simplified transfer function of the steam turbine with only high pressure section and disclaimer of crossover piping in comparison to reheater could be expressed:

$$W_{Ts}(s) = \frac{\Delta T_m}{\Delta V_{CV}} = \frac{1 + T_{RH} F_{HP} s}{(1 + T_{CH} s)(1 + T_{RH} s)} \quad (2)$$

It shows how turbine mechanical torque reacts to change of control valve position. *TRH*and *TCH* represent inertia time constant of the reheater and inlet steam chest while *FHP*shows the fraction of high steam pressure section. Also, T_m is turbine mechanical torque and *VCV* is control valve position.

The turbine control function is similar to hydraulic turbine. However in order to get stable operation of thermal power plant, it is enough to implement governor with a 4–5% speed droop. So there is no need of two types of droop compensation compare to hydro power plant. On the other hand thermal power plant output highly depends on primary fuel system and boiler operation and control. Fig. 3 illustrates the power output change of TPP in respect to control mode.

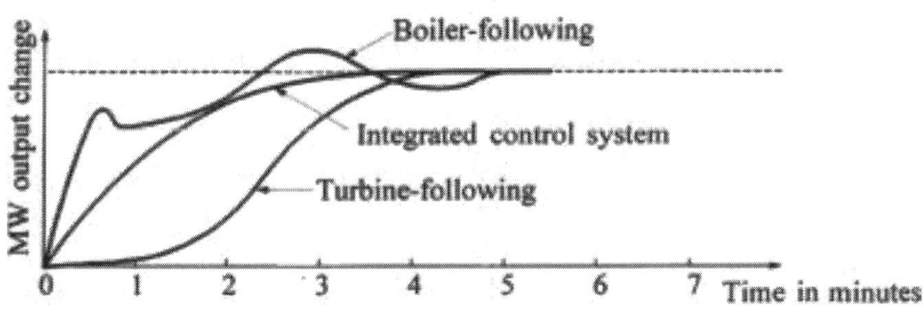

Figure 3. TPP power output in respect to control mode [29].

In the boiler following mode the turbine control valves initiates the changes in generation, while in turbine following mode the changes in generation are implemented by combustion controller [31].

The typical power outputs of steam and hydraulic turbines are shown in Fig. 4. Depending on the boiler type, the control mode and the size of the load change, the power output of thermal power plant night change significantly slower than illustrated. However hydraulic power plant output with a low head could be significantly faster than considered here.

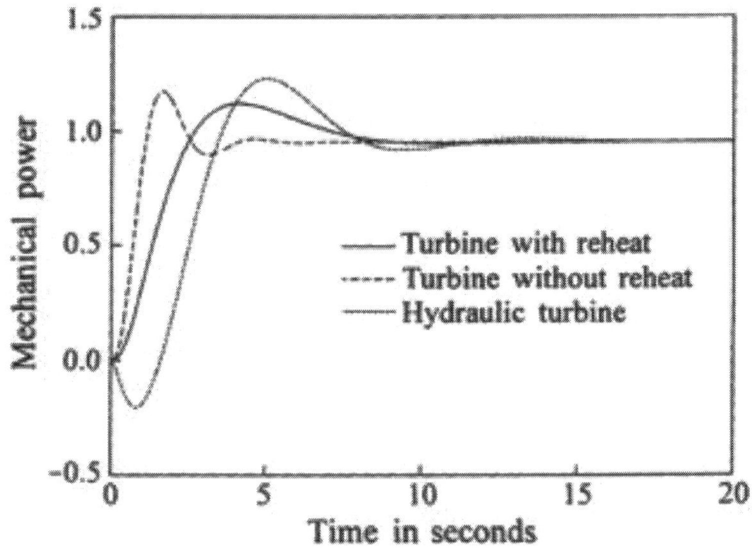

Figure 4. Typical responses of steam and hydraulic units [29].

Flow batteries, sometimes called redox batteries (i.e. oxidation and reduction reactions), are electrochemical systems, which are an alternative between the usual batteries and fuel cells [32]. Flow batteries could be charged as ordinary batteries and provide energy as long as charged electrolyte is supplied. The charging and discharging cycles are possible due to reversible electrochemical reaction between two electrolytes [33]. Conversely charged electrolyte is pumped through separate contours and reaction takes place in special ionic membrane as shown in Fig. 5[34].

The power of flow batteries depends on the surface area of the electrodes. It also depends on the quantity of the bi-pole electrodes. The capacity of flow battery could be increased by expanding the volume of the electrolyte reservoirs, thus increasing the amount of the electrolyte [35] and [36]. The modules of flow batteries are connected into groups in series to ensure the required voltage, but hydraulic circuits are connected in parallel in order to share electrolyte between groups, therefore the same charging level is ensured [37].

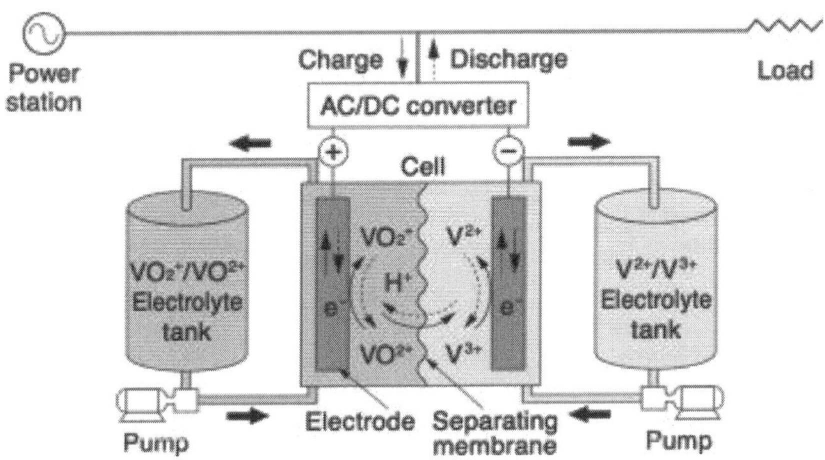

Figure 5. Principle and configuration of a flow battery [34].

Some advantages of flow batteries are that they can operate in low temperature and pressure conditions. Also, the electrolyte could be discharged completely [38]. The electrochemical processes are very fast so the reaction time of the batteries is rather small (0.04–0.06 s) and mostly depends on the operation of power electronics. Thus, flow batteries with power ratings in megawatt range could be very useful for power system balance. As it will be seen from the results of investigations the necessary power ratings of FB reaches tens of MW.

3. DYNAMIC MODELS OF HYBRID POWER SYSTEMS

MATLAB Simulink software environment was chosen to model and simulate the system. The goal of this research is to investigate the potential of flow battery technology to serve as energy balancing tool. TPP and HPP are used to cover low and mid-frequency imbalances respectively, whereas FB acts on high frequencies thus a trade-off between conventional power plant equipment wear and required battery size can be observed. It is important to determine the required flow battery parameters (power and capacity ratings) in order to maintain balance in the power system. During this research a model of TPP, HPP and FB was proposed with a control strategy.

3.1 Control strategy of hybrid power system

The proposed energy balancing method aims to reduce the imbalance of a virtually isolated electric power system. It consists of thermal power plant, hydro power plant, flow battery and a PI controller that mainly deals with the compensation of energy losses related to the flow battery charge/discharge cycle (Fig. 6). The model is designed to offer a tradeoff between equipment wear and the required size of the battery. As discussed later in the paper, bigger time constant in the low-pass filter (LPF) requires larger battery size and higher power ratings. The initial error is the difference between the actual power and the forecasted power:

$$P_{error} = P_{act} - P_f \tag{3}$$

where P_{error} is the initial error, P_{act} is actual power and P_f is forecasted power.

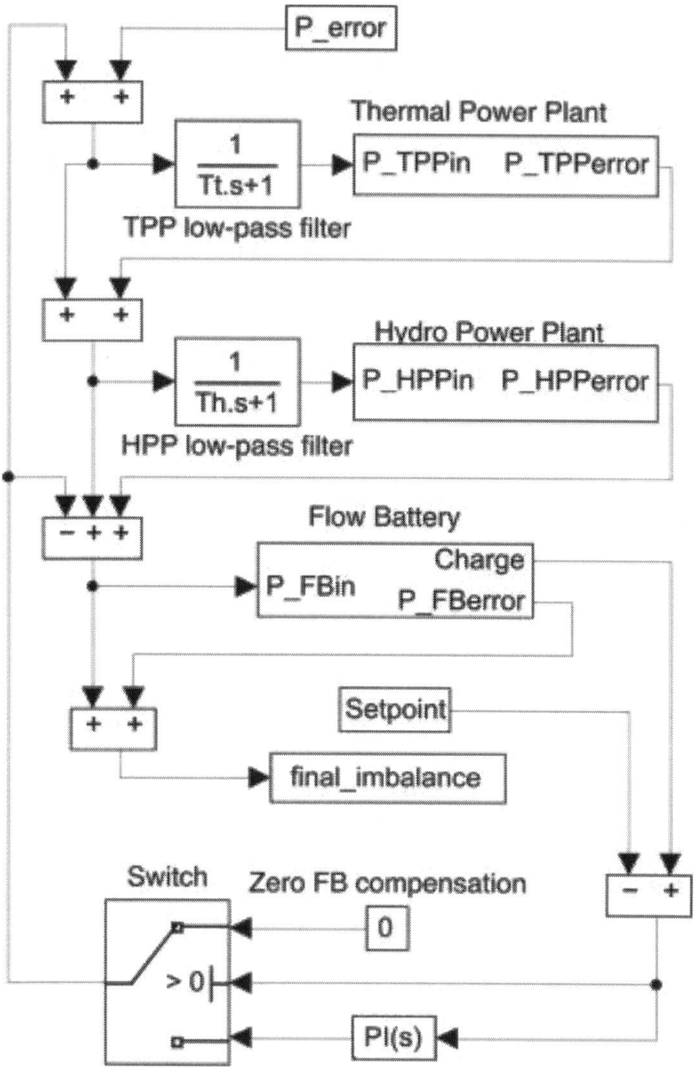

Figure 6. Model of the system.

The error between the actual and forecasted generation is first fed through rate limiting low-pass filter. It cuts off mid and high frequencies and reduces TPP depreciation costs. Next, the error left after TPP, is fed through another low-pass filter with a slightly lower time constant. This separates the mid-frequency band, which then goes to the HPP as a control input. It is clear that HPP changed its power output according to mid-frequency variations. This technique extends the lifetime of the HPP servo equipment. Generally HPP cannot change its power output very fast due to physical limitations, such as slew rate of the servomechanism and water inertia, which might cause water hammer. Thus, HPP responds by adjusting its power output to compensate only mid-frequency component.

On the other hand, the error that is left (mainly consisting of high frequencies) is then fed to a storage device – a flow battery. The small time constant and high charge/discharge power handles high frequency power fluctuations and smoothens the total power output from the system. Due to the fact that flow battery has a cycle efficiency of about 85% [39], additional energy to compensate energy losses is required. This is done by the feedback loop that signals TPP and HPP to overcome the losses. The signal adds nearly constant power compensation.

The proposed model is designed so that the average charge in the battery stays around half of its capacity. This deviation in battery charge from half of the capacity is fed to the PI controller as the error i.e. the desired battery charge set point is half the battery size. The controller responds by signalling TPP and HPP to adjust its generation to maintain the charge of FB at the desired level. Again, the charge in FB fluctuates but on average battery charge is kept constant. This level is proposed to be half the total capacity of the battery to be able to equally compensate both energy shortage and surplus. Also, a switch is added to compensate battery losses only when the battery's state of charge is below 50%. The operation of the battery is discussed in greater detail in Section 'The results of proposed balancing technique'.

To sum up, the frequency spectrum of the initial imbalance is divided into three bands – low, mid and high. The lowest frequencies are handled by TPP, mid-frequencies are cancelled by HPP and what is left – high frequencies – using a flow battery storage device.

3.2 Flow battery model

The main characteristics of flow batteries were estimated during the process of modelling the flow battery. The model does not take into account any electrochemical processes inside the cell nor the kinetic energy of the electrolyte itself. The main parameters considered were the limits of the power and energy (P*min*,P*max*,E*min*,E*max*), losses (efficiency) and reaction time. The main objective of the flow battery model was to simulate the response to power imbalance. When balancing wind power, the power to be balanced by the FB is determined as the difference between the initial error and power generated by TPP and HPP:

$$P_{FBin} = P_{error} - P_{TPP} - P_{-HPP} \tag{4}$$

where P*FBin* is the power to be balanced by the flow battery, P*error* is the initial power imbalance, P*TPP* and P*HPP* is power generated by thermal and hydro power plants correspondingly.

The main principle is to charge the battery when there is a surplus of energy and to discharge when the energy is scarce. This is depicted in Fig. 7. As it can be seen from the model in Fig. 8, power of flow battery should be kept within the interval [P*min*,P*max*], and energy stored in the flow battery E*FB* should stay within the limits of [E*min*,E*max*]. Controlling the flow battery's charge and discharge rate should compensate the high frequency part of the wind power variation from forecasted profile.

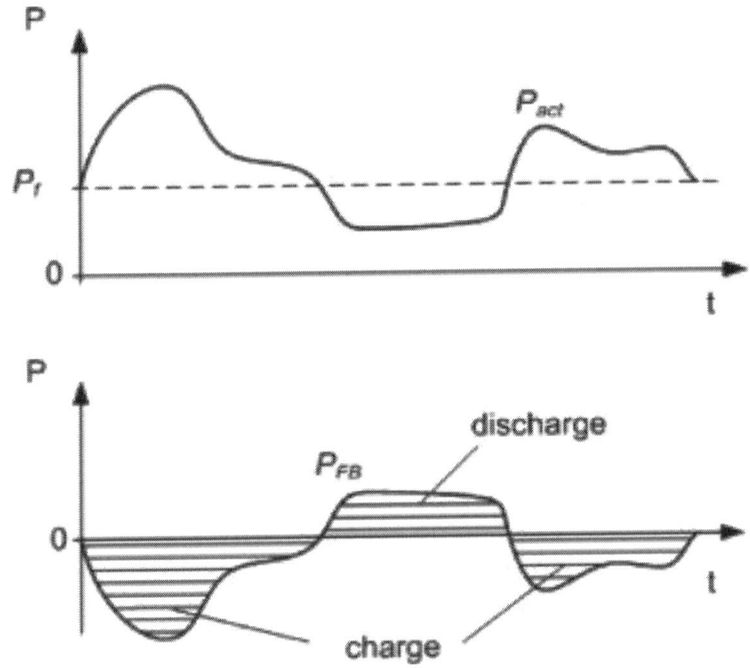

Figure 7. The control principle of the FB. P*act* – actual wind power, P_f – forecasted wind power, P*FB* – power of flow battery.

Also, the relative cycle losses are equally divided into charging and discharging losses using the following formula:

$$\eta_{one-way} = \sqrt{\eta_{cycle}}, \tag{5}$$

$$L_{one-way} = 1 - \eta_{one-way}, \tag{6}$$

$$L_{one-way} = 1 - \sqrt{\eta_{cycle}} \tag{7}$$

where η*one-way* is the efficiency of charging or discharging of the battery, η*cycle* is the total cycle efficiency and L*one-way* are the losses associated with either charging or discharging of the battery.

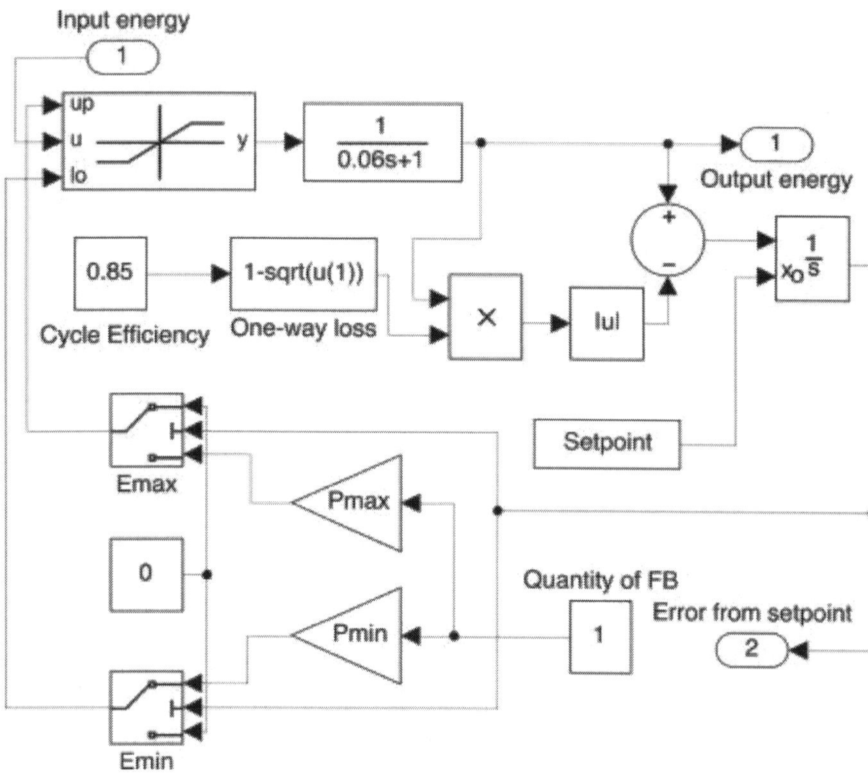

Figure 8. The structure of the flow battery model.

Some assumption were made while designing dynamic model of the flow battery:

- The total efficiency of the flow battery cycle is $\eta cycle$=85%[39],
- The inertia time constant is 0.06 s (considering inertia of power electronics) [40],
- The total discharge of the flow battery is allowed (E*min*=0) [38].

3.3 Model of hydro power plant

A hydro turbine was used to compensate the imbalance in the system that is left after TPP. The turbine output power follows the load variation trend and aims to reduce the error. It also helps to compensate losses associated with the FB (see Section 'Control strategy of hybrid power system'). The HPP was modelled in matlab using traditional governor controller/regulator popularly found in the literature and transfer functions [29]. In particular, a transfer function of hydro turbine is shown in Eq. (1) and the block diagram can be found in [30, p. 33]. The rest of the hydraulic turbine and speed regulator model parameters are given in Table 1. These parameters are chosen to match Kruonis hydro pumped storage power plant in Lithuania.

Table 1. Parameters of the hydro turbine.

Parameter	Notation	Value
Permanent droop	R	0.06
Temporary droop	r	0.5
Temporary droop time constant	T_r	5 s
Auxilary servo motor time constant	T_f	0.2 s
Gate servo motor time constant	T_g	0.2 s
Water time constant	T_w	4 s

The power to be balanced by the HPP is calculated using the following equation:

$$P_{HPPin} = P_{TPPerror} \times \frac{1}{T_{hLPF}S + 1} \tag{8}$$

where P*HPPin* is the power to be balanced by the HPP, P*TPPerror* is the power imbalance after TPP and the T*hLPF* is the time constant of the LPF related to hydro power plant.

3.4 Model of thermal power plant

Thermal transient process constrains and specific construction of thermal power plant could cause power to vary significantly slower than that of hydro power plant. It only follows the trend and does not reduce the error noticeably. Fig. 6 represents the whole system and contains TPP block. This block represents a general TPP model and has been created in Matlab according to model found in [29, p. 436].

The model of turbine consists of three cylinders: high pressure, intermediate and low pressure cylinders. Turbine is described with linear model and the transfer function of the turbine is:

$$W_T(s) = \frac{K_H(1 + sT_{CO})(1 + sT_R) + K_I(1 + sT_{CO} + K_L)}{(1 + sT_{SC})(1 + sT_R)(1 + sT_{CO})} \quad (9)$$

where K_H, K_I, K_L – the power of high, intermediate and low pressure cylinders per units; TSC, T_R, TCO – time constants of steam chest, reheater and the crossover between intermediate and low pressure cylinders. Speed governor is modelled as a periodic link of servomotor and power change speed rate limiter which holds the speed within $Vmax$,$Vmin$ values.

The linear mathematical model of the regulating processes of steam turbine speed governor, could be described as second order transfer function WSG:

$$W_{SG} = \frac{1}{(1 + sT_{RM})(1 + sT_{SM})} \quad (10)$$

where TRM – time constant of speed relay, TSM – time constant of servomotor.

A block diagram of steam turbine can be found in [29] and [31]. The parameters of the turbine that was used for the investigation are presented in Table 2.

Table 2. Parameters of the steam turbine.

Parameter	Notation	Value
Steam chest time constant	T_{SC}	0.25 s
Reheater time constant	T_R	5 s
Crossover piping time constant	T_{CO}	0.5 s
Factor of high pressure section	K_H	0.3
Factor of intermediate pressure section	K_I	0.3
Factor of low pressure section	K_L	0.4
Speed relay time constant	T_{SR}	0.1 s
Speed motor time constant	T_{SM}	0.3 s

Overall boiler model's transfer function W_B, as boiler's pressure p_B and fuel flow m*FL*ratio, when considering the constant steam mass flow from boiler:

$$W_B(s) = \frac{e^{-sT_D}}{T_B s(1 + sT_{FL})(1 + sT_W)} \tag{11}$$

where T_D – fuel feed delay time constant, T*FL* – heat transfer inertia time constant to pipes during fuel burning and T_W – inertia time constant of pipes for heat transfer to water and steam.

A block diagram from [31, p. 2] was used to create Matlab model. Boiler parameters used for the investigation are presented in Table 3.

Table 3. Boiler parameters.

Parameter	Notation	Value (s)
Fuel feed delay time constant	T_D	10
Boiler heat accumulation time constant	T_B	100
Heat transfer to pipes inertia time constant	T_{FL}	7
Heat transfer to water and steam time constant	T_{WP}	6

The power to be balanced by the TPP is calculated using the following equation:

$$P_{TPPin} = (P_{error} + PI_{out}) \times \frac{1}{T_{tLPF}s + 1} \quad (12)$$

where P*TPPin* is the power to be balanced by the TPP, P*error* is the initial power imbalance, *PIout* is the output from the PI controller and the T*tLPF* is the time constant of the LPF related to TPP.

4. INVESTIGATION OF WIND POWER IMBALANCE

The actual data of wind farms installed in Lithuania was used in this research (Fig. 9). In this particular case, it is the forecasted data and the actual wind power data for the period of 1st to 10th of February 2014. The total installed capacity at that date was 222 MW. The mean absolute percentage error (MAPE) of the forecasted wind power during the investigated period was 32% with a standard deviation (SD) of 61 MW. A histogram of the initial errors can be seen in the results section. It should be noted that this period was chosen due to high variation in wind generation as well as high prediction mismatch in order to investigate wind balancing technique in extreme case.

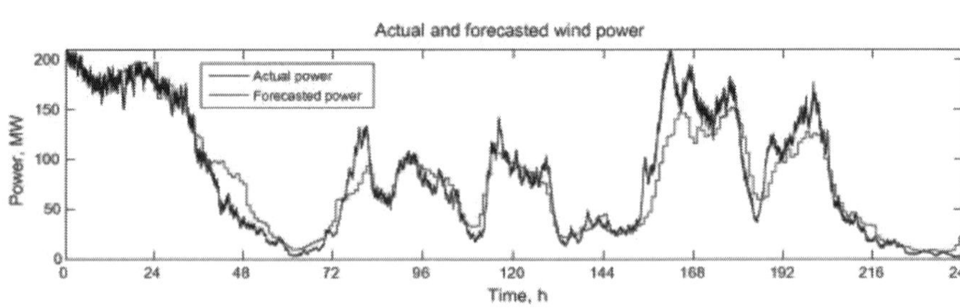

Figure 9. The forecasted and actual wind power for a installed capacity of 222 MW from 1st to 10th of February, 2014.

5. THE SENSITIVITY ANALYSIS OF LOW-PASS FILTER CUT-OFF FREQUENCIES

In order to investigate the influence of the low-pass filters cut-off frequencies to the proposed balancing system, a sensitivity analysis of the low-pass filters time constants have been prepared. Thermal power plant low-pass filter time constant ranging from 0 to 10,000 s (step of 500 s) and hydro power plant low-pass filter time constant of 0–500 s (step of 25 s) have been tested. The main parameters of the proposed system, such as the required active power reserve and mean power rates of thermal and hydro power plants, also the required capacity and power ratings of flow battery have then been analysed.

The dependence of required TPP active power reserve, for different low-pass filter cut-off frequencies could be seen in Fig. 10, while mean power is shown in Fig. 11. The mean power of TPP could represent the total energy generated by the power plant as a regulating energy. It can be clearly seen from the figures that higher cut-off frequencies (lower low-pass filters time constants) increase active power reserve as well as regulating energy of the thermal power plant. It could also be noted that low-pass filter time constant of hydro power plant has minor influence on the thermal power plant specific parameters.

Different situation could be identified in respect to HPP specific parameters. Both the HPP low-pass filter time constant and TPP low-pass filter time constants have appreciable influence. The required active power reserve of HPP is shown in Fig. 12, whereas Fig. 13 represents mean power of this type of plant.

Active power reserve of HPP increases when TPP low pas-filter parameter increases and HPP low pass-filter parameter decreases. It could be seen in Fig. 12.

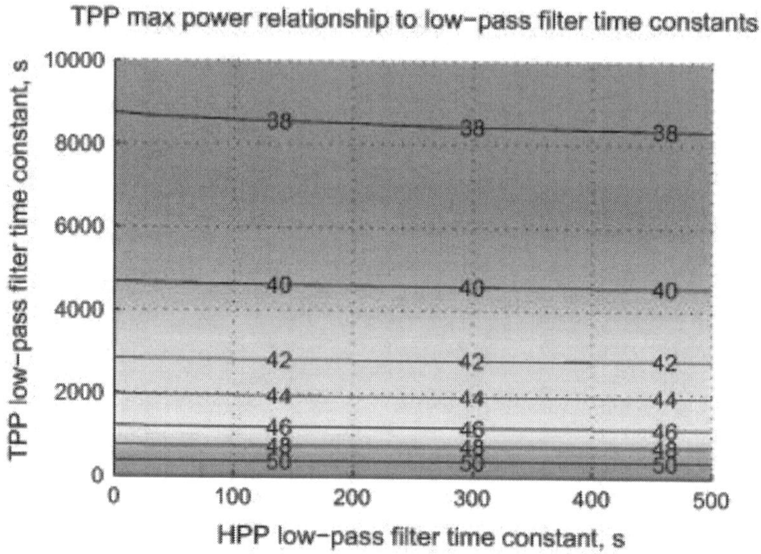

Figure 10. Relationship of required TPP active power reserve with respect to low-pass filter time constants.

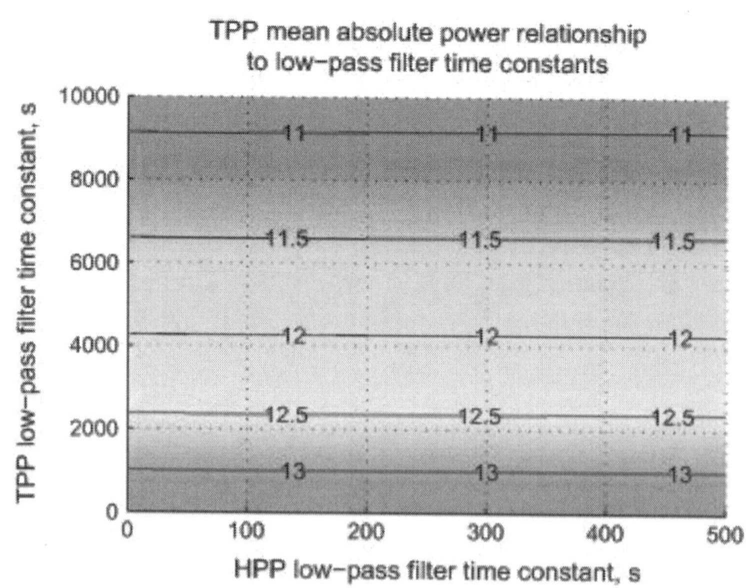

Figure 11. TPP mean absolute power relationship with respect to low-pass filter time constants.

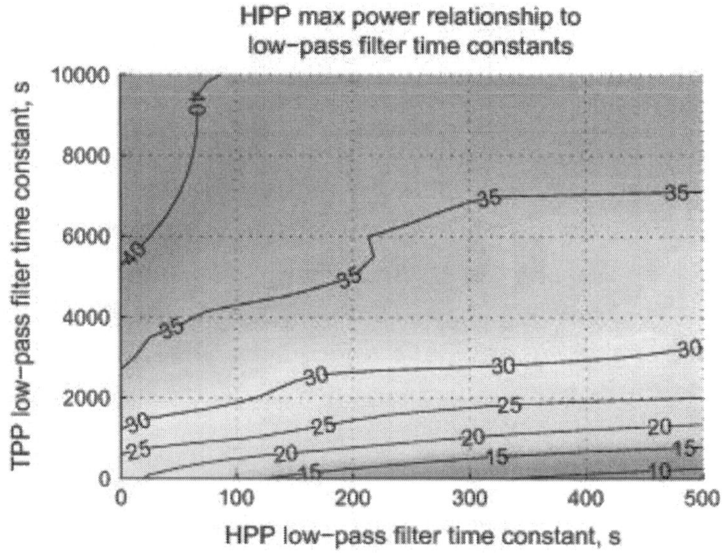

Figure 12. Relationship of required HPP active power reserve with respect to low-pass filter time constants.

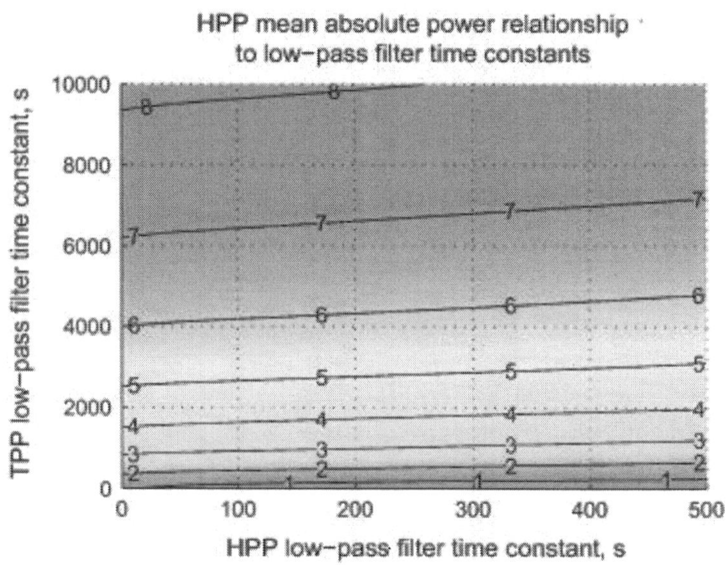

Figure 13. HPP mean absolute power relationship with respect to low-pass filter time constants.

A series of simulations of different low-pass filters time constants have also been run in order to get the flow battery specific parameters, in relationship with different cut-off frequencies. The required flow battery capacity and power range are represented in Fig. 14 and Fig. 15 respectively. Flow battery required parameters, as shown in the figures, depend on both low-pass filters' time constants. But the main influence is done by the one associated with HPP. As the parameter of HPP low-pass filter time constant increases, the flow battery specifications also increase. It can be noticed that the flow battery mean absolute power relationship to low-pass filters cut-off frequencies is also similar, and taken into account for the analysis.

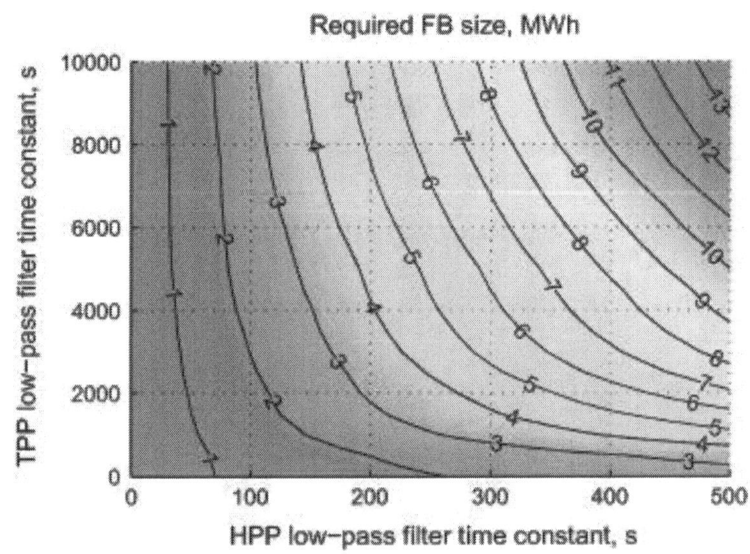

Figure 14. Relationship of required flow battery capacity with respect to low-pass filter time constants.

One of the most important things in this sensitivity analysis is to identify the relationships of thermal and hydro power plants' active power rates, which represent the intensity of power output variation of these power plants as well as the asset depreciation. Thermal power plant mean absolute power rate is represented in Fig. 16.

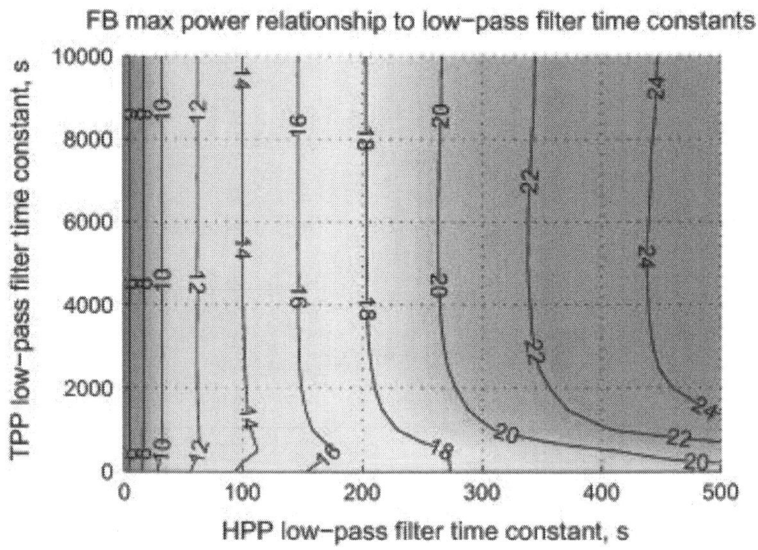

Figure 15. Relationship of required FB active power reserve with respect to low-pass filter time constants.

The significant mean absolute power rate downfall was identified when thermal low-pass filter time constant increase. In order to show the better visibility the low-pass filer time constant of HPP was kept constant at 75 s. However the full relationship on both filter parameters was assessed in the investigation.

Mean absolute power rate of HPP is shown in Fig. 17. Similar results to TPP mean power rate relationship can be observed. A significant drop was identified when hydro power plant's low-pass filter parameter increases. The TPP low-pass filter time constant was also kept constant at 5000 s due to better visualisation of the dependency, while the relationship on both filter parameters was taken into account in this analysis.

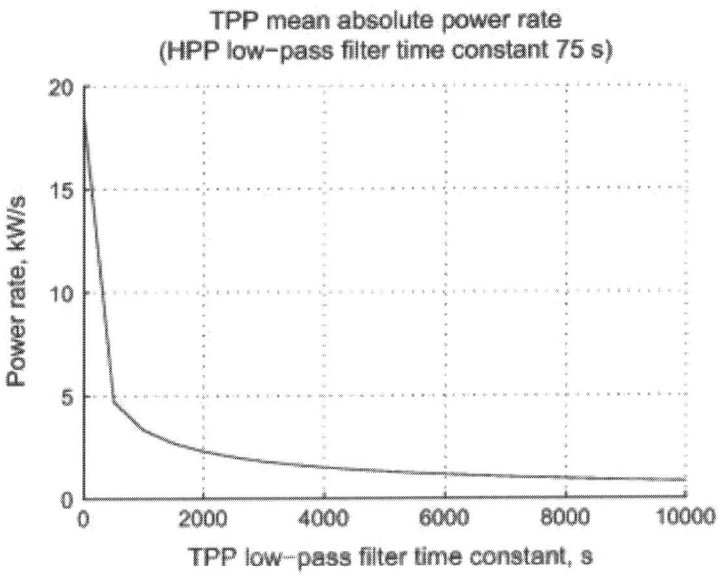

Figure 16. TPP mean absolute power rate with respect to low-pass filter time constant.

6. OPTIMAL LOW-PASS FILTER PARAMETERS IDENTIFICATION

Sensitivity analysis presented in the previous section has shown various dependencies of model parameters and relationships. It was identified that required active power reserve in conventional power plant increases as low-pass filter time constant decreases. However specific parameters of the flow battery have opposite relationship as well as active power rates and intensity of traditional power plants regulation. Optimal low-pass filters time constants and cut-off frequencies are investigated in this section.

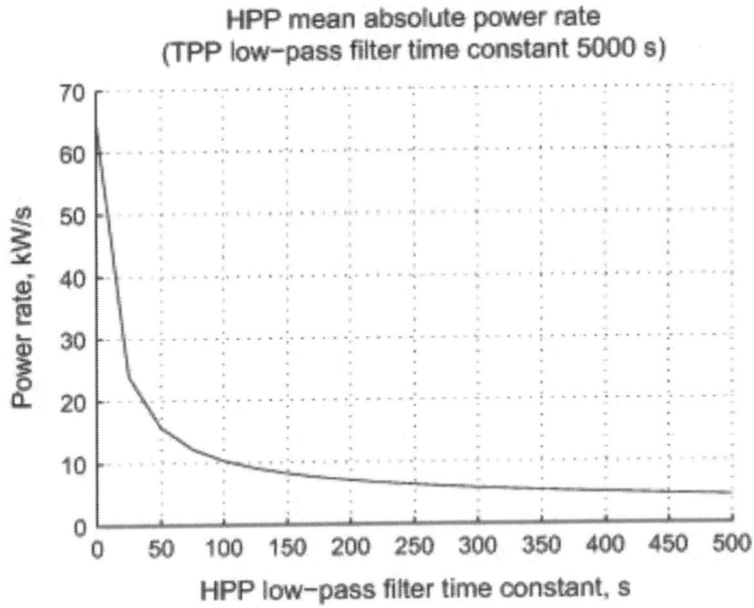

Figure 17. HPP mean absolute power rate with respect to low-pass filter time constant.

Objective function was prepared in order to investigate the optimal parameters and the best operation of proposed balancing system:

$$\begin{aligned} min \quad C = & C_1 \times P_{TPPmax} + C_2 \times P_{HPPmax} + \\ & C_3 \times P_{TPPmean} + C_4 \times P_{HPPmean} + \\ & C_5 \times R_{TPP} + C_6 \times R_{HPP} + \\ & C_7 \times P_{FBmean} + C_8 \times P_{FBmax} \\ & C_9 \times E_{FB} \end{aligned} \quad (13)$$

where P*TPPmax* and P*HPPmax* are required active power reserve of conventional power plants, P*TPPmean* and P*HPPmean* represents mean absolute power generated by thermal and hydro power plants, R*TPP* and R*HPP* are active power rates, P*FBmean*,P*FBmax* and E*FB* are specific parameters of flow battery – active mean power, active power range and required capacity respectively. Finally $c_1, c_2, c_3, \ldots, c_9$ are relative price corresponding to each of previously mentioned parameters. Different price ratios were used in order to obtain optimal parameters and to avoid conflicts as the price is always controversial. Table 4 shows specific relative prices which were used in this investigation. The minimisation of objective function leads to identification of the optimal low-pass filter parameters and overall operation of proposed balancing system.

Table 4. Relative prices.

Parameter	Price notation	Value
TPP reserve power	C_1	0.03
HPP reserve power	C_2	0.02
TPP energy	C_3	0.1
HPP energy	C_4	0.15
TPP mean absolute power rate	C_5	350
HPP mean absolute power rate	C_6	30
FB mean power	C_7	0.005
FB max power	C_8	0.05
FB capacity parameter	C_9	0.1

Results of low-pass filters time constants objective function are represented in Fig. 18. The region of the minimum objective function could be seen with the parameters – TPP low-pass filter time constant 5000 s and HPP low-pass filter time constant 75 s. These values will be used in time domain simulations of wind power balancing.

The relationship of the low-pass filter parameters could be expressed:

$$f_c = \frac{1}{2\pi\tau} \quad (14)$$

where f_c is the filter cut-off frequency and τ is the filter time constant.

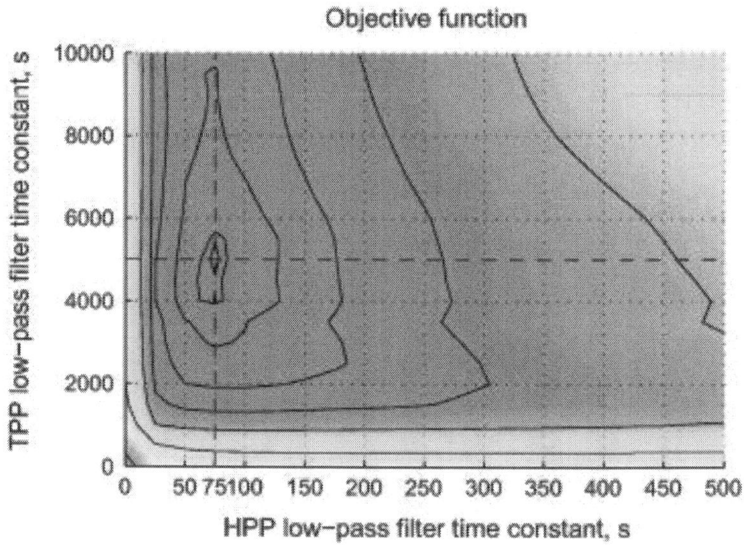

Figure18. The output of the objective function. A minimum can be seen then time constants are 75 and 5000 for HPP and TPP respectively.

Fig. 19 shows the initial imbalance of wind power and conventional power plants output generation as well as flow battery output in respect to signal frequency decomposition using identified optimal low-pass filter parameters. It can be clearly seen that low frequencies of initial wind power imbalance are covered by thermal power plant, while hydro power plant deals with mid-frequencies and the flow battery eliminates high frequencies.

7. THE RESULTS OF PROPOSED BALANCING TECHNIQUE

The simulations were run using the chosen optimal time constants of 5000 s and 75 s for TPP and HPP correspondingly and the data from Fig. 9. The results are shown in Fig. 20. It includes the initial power system imbalance, difference after both TPP and HPP and final imbalance as well as the operating power of TPP, HPP and FB. After the addition of

TPP, the initial imbalance SD dropped from 19.5 MW to 9.2 MW. Similarly after the addition of HPP, system imbalance improved to SD of 1.65 MW. Finally, after FB the SD of imbalances were 0.1 MW. Fig. 21 shows the histogram of initial imbalances in green, imbalances after the TPP in red and imbalances after HPP in blue. From the same graph it can be observed how the spread of imbalances decrease while propagating through the system. It should be noted that addition of TPP and HPP gave positive results and, the mean over-generation of TPP and HPP is equal to mean losses in FB (about 80 kW). This justifies the validity of the model.

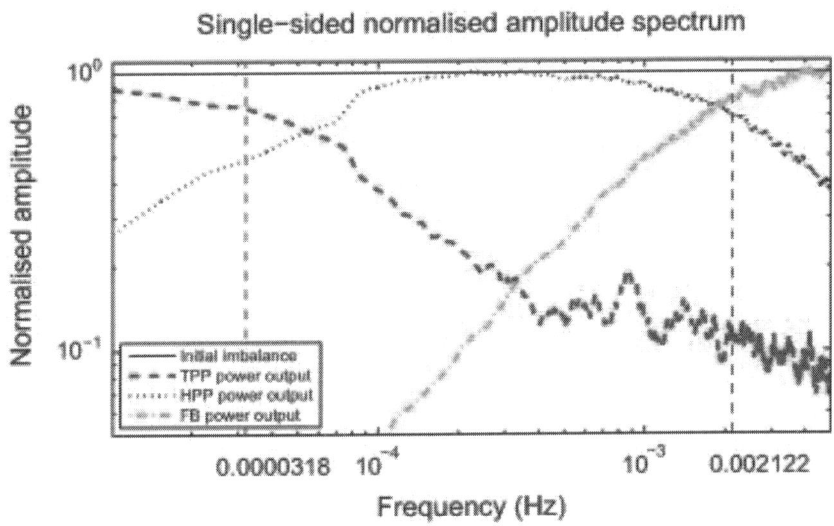

Figure 19. Frequency decomposition of initial wind power and TPP, HPP, FB output powers.

The top part of Fig. 20 shows a three day simulation period. This period is enough to demonstrate the performance of the system and the nature of TPP and HPP output. The middle part of the diagram also shows a magnified portion of errors (between the time of 12:00 and 13:00 h of the first day) and, it can be seen how the HPP power tracks the trend of the imbalance after TPP. On the other hand, the bottom part of the graph shows how FB follows the imbalance after HPP and the graph in red is the final imbalance (time from 12:00 to 12:30 of the first day). The MAPE has now dropped to 0.068%.

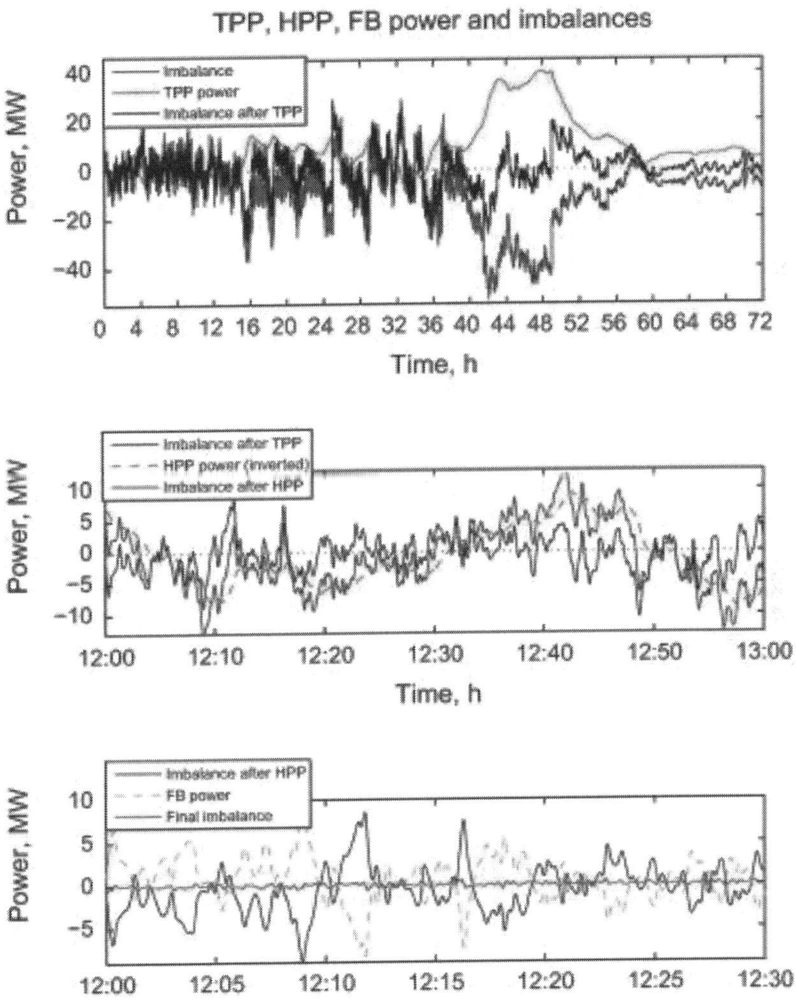

Figure 20. Balancing the difference between actual and forecasted wind power with thermal power plant, hydro power plant and flow battery.

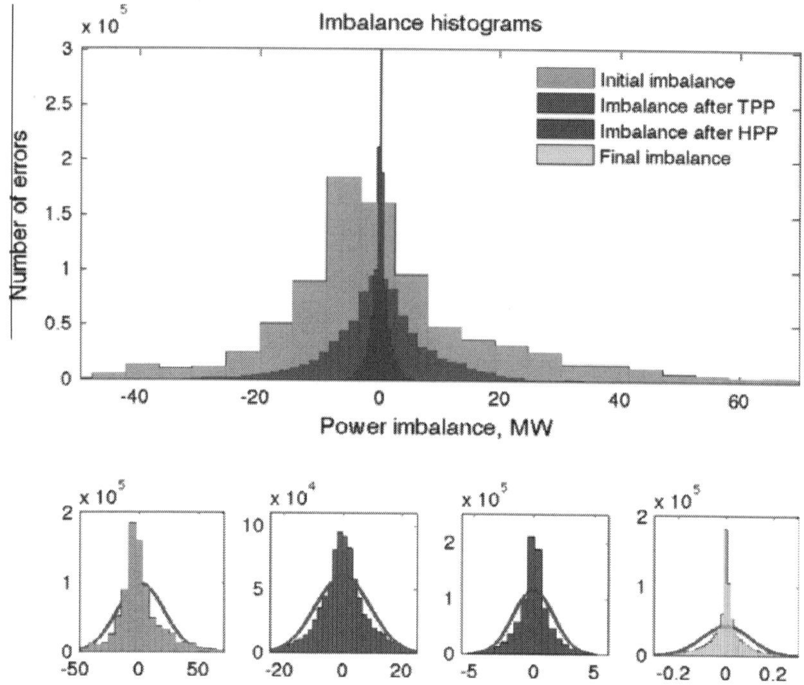

Figure 21. Histograms of imbalances at different stages in the system.

During the investigation, the required power and energy of the flow battery was determined to be 55 MW and 1.90 MW h respectively. This is about 66.4% of the mean wind power during the investigated period and about 24.9% of the total installed wind power capacity in the Lithuanian power system (the total wind power is about 222 MW). A more detailed discussion on why the required FB ratings might be chosen lower can be found in Section 'Limiting the maximum required FB power'. Overall, these results can be considered as feasible for implementation.

Fig. 22 shows the charge level and accumulative losses in the FB during the simulated period. The required battery size is recorded to be 1.90 MW h. This is the total required capacity in a perfect balance situation. If some power and energy spikes were ignored, the required battery power and size would be considerably lower. To cover the spikes, a super capacitor could potentially be used to serve the required ultra short term power.

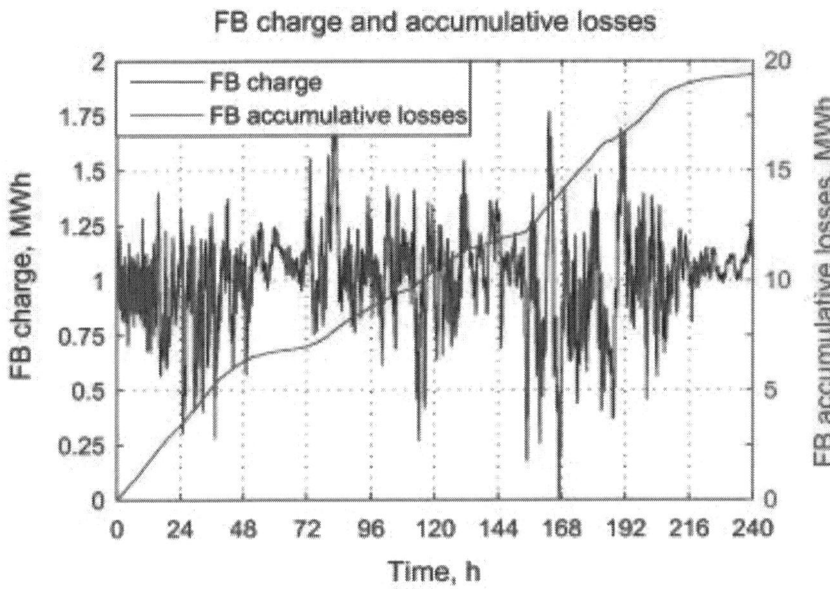

Figure 22. FB charge and accumulative losses.

8. LIMITING THE MAXIMUM REQUIRED FB POWER

Considering the SD of power to be balanced by the FB and the three-sigma rule it is reasonable to limit the installed FB power because high power is required during a negligible amount of time. This section discusses the trade-off between FB power rating and the final balance of power.

The data from simulations was extracted to calculate the balanced portion of energy depending on battery power and low-pass filter time constants (Fig. 23). The graph shows the FB power and capacity ratings needed to balance different portions of initial energy.

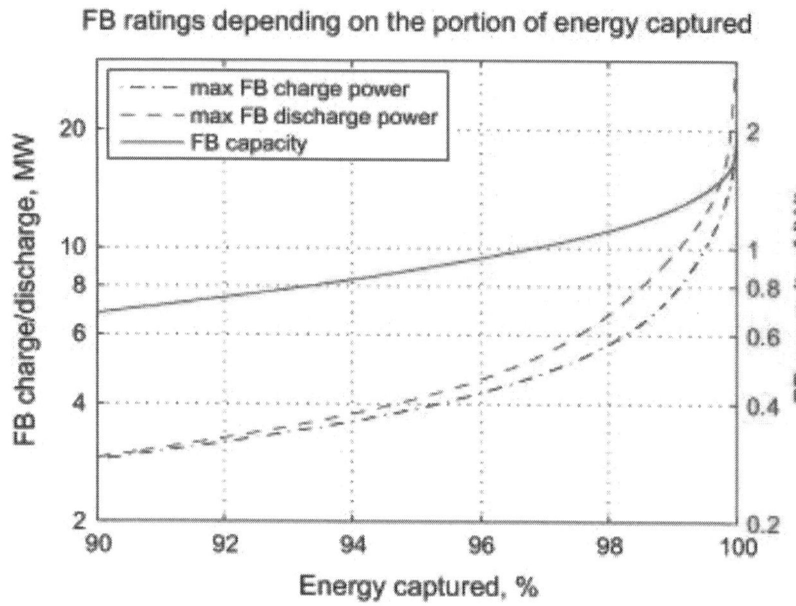

Figure 23. Relative balance dependancy on FB rated discharge and LPF time constant.

As it can be seen from Fig. 23, when 99.7% of energy is balanced, the required FB maximum power is about 14.1 MW. That is a significant drop in the required FB power rating. The required capacity of the FB would also become 1.47 MW h. On the other hand, this limits the power needed for balancing, which results in the increased final imbalance MAPE to 0.073% and a SD to 0.2 MW, but is still quite small compare to initial values (32% and 19.5 MW).

A further decrease in energy capture could be considered. The 95% would correspond to 4.1 MW FB power limit and 0.88 MW h required FB energy and would increase MAPE of final imbalance to 0.14% and SD to 0.58 MW, but is still 43% amount of imbalance improvement compared to the value without hybrid system operation.

9. CONCLUSIONS

The proposed hybrid wind power balancing technique, using TPP, HPP and FB control strategy, presented positive results in balancing the wind power. It also generalises a technique to find the optimal ratings for a flow battery in the Lithuanian power system. However this method could be applied to other electric power systems as well. It might be particularly important to power systems, which operate in island mode.

In this paper, a series of simulations were carried out to identify the cut-off frequencies for the low-pass filters, which optimally controls the power output of thermal and hydro power plants. The optimal cut-off frequency identification enabled the estimation of the required flow battery power and capacity ratings. Also, the paper mainly focuses on the technical side of the method instead of looking at the economical value in detail.

After implementing the proposed control strategy for the TPP, the initial imbalance decreased by about 24% (from 32% MAPE to 24.36%). A further imbalance decrease was reached after the HPP was added – from 24.36% MAPE to 1.44%. After introducing flow batteries, the system became fully balanced. However, it required FB size of 1.9 MW h and power of 55 MW. This constitute to about 25% of the total installed wind power in the Lithuanian power system. Most of the FB's power and capacity is required when the power system is in emergency state. It should be mentioned that such ratings were required when covering every moment in time.

On the other hand, these events are relatively rare, besides some types of flow batteries tolerate overloads for short periods without negative side effects. Making such assumptions the results of additional investigation showed that more reasonable flow battery ratings could be chosen. By covering 99.7–95% of the power imbalance, it is possible to reduce FB power rating by about 4–13 times and FB required capacity from about 23–54% (Table 5 and Table 6).

Table 5. Overview of results (Imbalances).

Imbalance stage	MAPE (%)	SD (MW)	ME (kW)
Initial imbalance	32.00	19.52	1291
After TPP	24.36	9.20	−78.9
After HPP	1.44	1.65	−81.5
100% final	0.068	0.10	−0.02
99.7% final	0.073	0.20	1.6
95% final	0.144	0.58	6.3

Table 6.. Final balance level vs. required FB ratings.

Balance level (%)	FB rated power (MW)	FB capacity (MW h)
100	55.3	1.90
99.7	14.1	1.41
95	4.1	0.88

Many other storage technologies could be similarly modelled and investigated. In particular, high power and low capacity storage devices, such as super capacitors, could be added to compensate highest frequency imbalances thus highly improving results and reducing power requirements for the FB. Having many different power plants in the model it is then potentially useful to research control strategies in order to reach for the highest economical or environmental benefit.

ACKNOWLEDGEMENTS

The authors would like to acknowledge the funding support from EPSRC via Faculty of Science and Technology, Lancaster University, UK. The authors would also like to thank the Department of Power Systems, Kaunas University of Technology, Lithuania and National Transmission System Operator in Lithuania for providing the necessary data. The data generated in this work are available from the Lancaster University data archive[41].

REFERENCES

1. Gelazanskas L, Gamage KAA. Demand side management in smart grid: a review and proposals for future direction. Sustainable Cities and Society.

2. Le HT, Santoso S, Nguyen TQ. Augmenting wind power penetration and grid voltage stability limits using ESS: application design, sizing, and a case study. IEEE Trans Power Syst 2012;27(1):161–71.

3. Diaz-Gonzalez F, Sumper A, Gomis-Bellmunt O, Bianchi FD. Energy management of flywheel-based energy storage device for wind power smoothing. Appl Energy 2013;110(0):207–19.

4. Tewari S, Mohan N. Value of NAS energy storage toward integrating wind: results from the wind to battery project. IEEE Trans Power Syst 2013;28 (1):532–41.

5. Union for the coordination of the transmission of electricity. Operation Handbook; 2013.

6. Denholm P, Ela E, Kirby B, Milligan M. The role of energy storage with renewable electricity generation. Tech. Rep. NREL/TP-6A2-47187. A national laboratory of the U.S. Department of Energy; January 2010.

7. Droste-Franke B, Paal BP, Rehtanz C, Sauer DU, Schneider J-P, Schreurs M. Balancing renewable electricity: balancing renewable electricity energy storage, demand side management and network extension from perspective an interdisciplinary; 2011.

8. Han X, Chen F, Cui X, Li Y, Li X. A power smoothing control strategy and optimized allocation of battery capacity based on hybrid storage energy technology. Energies 2012;5(5):1593–612.

9. Soder L, Hamon C. Power balance regulation at large amounts of wind power. Tech. rep., Elforsk.

10. Lubosny Z, Bialek J. Supervisory control of a wind farm. IEEE Trans Power Syst 2007;22(3):985–94

11. Jiang Q, Wang H. Two-time-scale coordination control for a battery energy storage system to mitigate wind power fluctuations. IEEE Trans Energy Convers 2013;28(1):52–61.

12. Abbey C, Strunz K, Joos G. A knowledge-based approach for control of twolevel energy storage for wind energy systems. IEEE Trans Energy Convers 2009;24(2):539–47.

13. Wang P, Gao Z, Bertling L. Operational adequacy studies of power systems with wind farms and energy storages. IEEE Trans Power Syst 2012;27 (4):2377–84.

14. Khalid M, Savkin A. Minimization and control of battery energy storage for wind power smoothing: aggregated, distributed and semi-distributed storage. Renew Energy 2014;64(0):105–12.

15. Khalid M, Savkin A. A model predictive control approach to the problem of wind power smoothing with controlled battery storage. Renew Energy 2010;35(7):1520–6.

16. Yuan Y, Zhang X, Ju P, Qian K, Fu Z. Applications of battery energy storage system for wind power dispatchability purpose. Electric

Power Syst Res 2012;93(0):54–60. http://dx.doi.org/ 10.1016/ j.epsr.2012.07.008. .

17. Ansari MMT, Velusami S. DMLHFLC (dual mode linguistic hedge fuzzy logic controller) for an isolated wind-diesel hybrid power system with BES (battery energy storage) unit. Energy 2010;35(9):3827–37.

18. Kalantar M, Mousavi G SM. Dynamic behavior of a stand-alone hybrid power generation system of wind turbine, microturbine, solar array and battery storage. Appl Energy 2010;87(10):3051–64.

19. Mousavi G SM. An autonomous hybrid energy system of wind/tidal/ microturbine/battery storage. Int J Electrical Power Energy Syst 2012;43 (1):1144–54.

20. Ma T, Yang H, Lu L. A feasibility study of a stand-alone hybrid solar/ wind/battery system for a remote island. Appl Energy 2014; 121(0):149–58.

21. Apostolopoulou D, Sauer PW, Domingeuz-Garcia AD. Automatic generation control and its implementation in real time; 2014.

22. Ram P, Jha AN. Automatic generation control of interconnected hydro-thermal system in deregulated environment considering generation rate constraints; 2010.

23. Gandhi NF, Mohan YK, Rao AV. Load frequency control of interconnected power system in deregulated environment considering generation rate constrains; 2012.

24. Cheng Y, Sahni M. Alti-ees automatic generation control study. Tech. rep., PwrSolutions Inc.; February 2012.

25. Bevrani H, Hiyama T. Intelligent automatic generation control. CRC Press; 2011. [26] Dagdougui H, Minciardi R, Ouammi A, Robba M, Sacile R. A dynamic decision model for the real-time control of

hybrid renewable energy production systems. IEEE Syst J 2010;4(3):323–33.

26. Yoshimoto K, Nanahara T, Koshimizu G. New control method for regulating state-of-charge of a battery in hybrid wind power/battery energy storage system. In: Power systems conference and exposition, 2006. PSCE '06. 2006 IEEE PES, 2006. p. 1244–51.

27. Castronuovo E, Peas Lopes J. On the optimization of the daily operation of a wind-hydro power plant. IEEE Trans Power Syst 2004;19(3):1599–606.

28. Kundur P. Power system stability and control. Reading, MA: McGraw-Hill; 1994. [30] Tenorio LAL. Hydro turbine and governor modelling. Norwegian University of Science and Technology. .

29. Gao L, Dai Y. Modeling large modern fossil-fueled steam-electric power plant and its coordinated control system for power system dynamic analysis. In: International conference on power system technology, 2010. POWERCON 2010.

30. Mohamed MR, Sharkh SM, Ahmad H, Seman MNA, Walsh FC. Design and development of unit cell and system for vanadium redox flow batteries (VRFB). Int J Phys Sci 2012;7(7):1010–24. .

31. Li L, Kim S, Wang W, Vijayakumar M, Nie Z, Chen B, Zhang J, et al. A new vanadium redox flow batteru using mixed acid electrolytes.

32. Shigematsu T. Redox flow battery for energy storage. Tech. Rep. 73, Sumitomo Electric Industries; October 2011. .

33. Weber AZ, Mench MM, Meyers JP, Ross PN, Gostick JT, Liu Q. Redox flow batteries: a review. J Appl Electrochem 2011;41(10):1137–64. .

34. Barote L, Weissbach R, Teodorescu R, Marinescu C, Cirstea M. Technologies for energy storage – present and future: flow batteries; 2008. p. 407–12.

35. Tokuda N, Kanno T, Hara T, Shigematsu T, Tsutsui Y, Ikeuchi A, Itou T, et al. Development of redox flow battery system. Tech. Rep. 50, Sumitomo Electric Industries; June 2000. .

36. Youa D, Zhanga H, Chen J. A simple model for the vanadium redox battery. [39] Teleke S, Baran ME, Huang AQ, Bhattacharya S, Anderson L. Control strategies for battery energy storage for wind farm dispatching, vol. 24; 2009. p. 725–32.

37. Connolly D. An investigation into the energy storage technologies available, for the integration of alternative generation techniques. Tech. rep., Department of Physics, University of Limerick; November 2007. [41] Gelazanskas L, Gamage K. Wind generation, 2015.

Chapter 9

Study of Performance of Fresnel Lens Solar Concentrator

R. R. Udawant , K. C. Mohite , M. G. Takwale

School of Energy Studies, University of Pune, Pune, India

ABSTRACT

The demand for process heat particularly low pressure steam is increasing in recent past. Solar energy can be utilized for generating medium temperature (100°C -150°C) low pressure (T<200°C) steam. By using solar concentrators, low pressure steam can be generated by using low cost technology [1]. Fresnel lens solar concentrators are very useful to generate low pressure steam for industrial and cooking applications. The study of performance of Fresnel lens solar concentrator and steam generation using Fresnel lens solar concentrator is presented in this paper. The overall efficiency of Fresnel lens solar concentrator is around 51% and can generate process steam required in the industry. The thermal performance and cost benefits of Fresnel lens solar concentrator are also described. The studies presented in this paper concerns with Fresnel Lens Solar Concentrator System which are most useful for generating process heat for industrial applications. Thermal performance of Fresnel lens solar concentrator

(FLSC) system can be studied by evaluating useful heat gain from the system when exposed to solar radiation [1]. FLSC system consists of Fresnel lens, receiver tracking system and heat removing device. An attempt has been made to evaluate the performance of FLSC system with two methods viz.continuous fluid flow test method and Steam generation test method. The study reveals that Fresnel lens Solar Concentrator system has a very good potential in generation of low pressure steam.

KEYWORDS

Solar Concentrator, Fresnel Lens, Solar Energy, Steamgeneration

1. INTRODUCTION

The Energy demand over the world is increasing exponentially, while fossil fuel reserves have started depleting. In order to sustain the development, continuous supply of energy is required. There is need to find energy resources which can be regenerated, such as solar energy, geothermal, biomass or will not deplete with time. Renewable energy source is one which can be regenerated or does not deplete with time. Solar energy is the most important source of energy which can fulfill future energy needs. There are many technologies developed to use solar energy for domestic use.

The Fresnel lens solar concentrator developed at school of energy studies, university of pune is designed for direct steam generation using solar energy. The solar concentrator system has been designed for generation of low pressure steam for various applications such as industrial process heat and steam cooking. The point focus and line focus Solar Concentrators are already been developed by industries in India and abroad [6, 7]. The glass and aluminium material are normally been used as reflectors in the solar concentrators. An unique attempt has been made to develop line focus refracting type solar concentrators for thermal applications. A procedure is developed for thermal performance evaluation of Fresnel lens solar concentrator for continuous fluid flow and direct steam generation [4]. The Fresnel lens solar concentrator developed at school of energy studies, university of pune is designed with following objectives.

Objectives

1. To estimate thermal performance of Fresnel lens solar concentrator.

2. To study the effect of mass flow rate, intensity of solar radiation and inlet temperature on efficiency of FLSC system.

3. To estimate quantity and quality of steam.

4. To estimate the efficiency of Fresnel lens solar concentrator at different operating pressures (steam temperatures).

2. EXPERIMENTAL

2.1. Continuous Fluid Flow Method

A setup for "continues fluid flow" test was developed in the laboratory, the schematic view of the test set-up is shown in the Fig.1 it essentially consists of,

1) Refracting type Fresnel lens and a receiver placed at the focal line of the lens.

2) Single axis tracking system.

3) Fluid circulation system consisting of storage tank, fluid circulation pump. Flow meter for heating inlet fluid to desired temperature and well insulated connecting pipe lens.

4) Two temperature sensors, one at inlet and other at the outlet of the receiver.

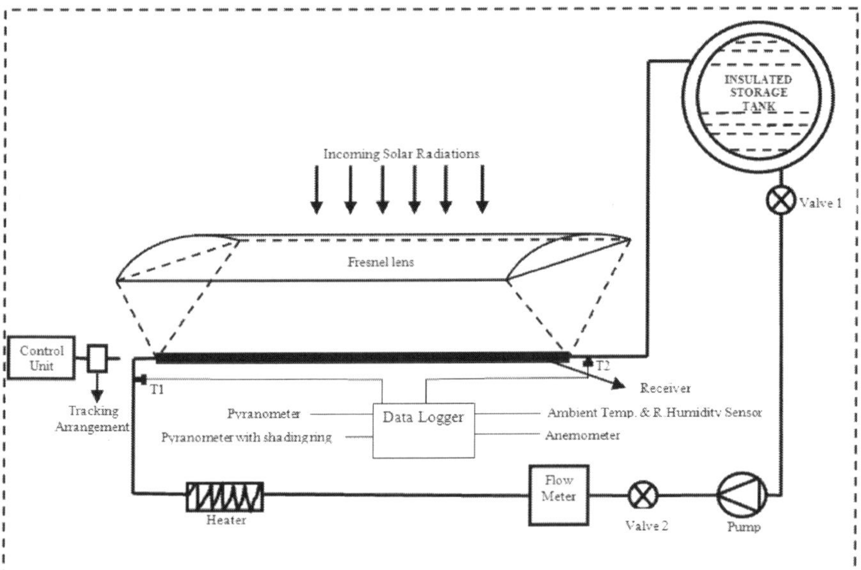

Figure 1. FLSC System Experimental setup (Continuous Fluid Flow)

In addition to FLSC system, following equipment's are used for collecting thermal performance data;

1) Pyranometer (1) for measuring Global solar radiation on horizontal plane.

2) Pyranometer (2) with shading ring to measure diffuse solar radiation.

3) Temperature and humidity sensor for measuring ambient temperature and relative humidity.

4) Anemometer for measuring wind velocity.

Temperature sensors shown in Fig.1 and above equipments are connected to data logger having 16 channels and memory module to collect and store the data.The protocol to evaluate thermal performance of FLSC system is essentially based on ISO standards. The experiments are carried out during clear sunny days and when intensity of solar radiation (beam) is more than 400 W/m^2. Following parameters are measured at the interval of 05 minutes. Data logger records average value of each parameter over a period of 05 minutes.

1) Inlet temperature of fluid (°C)

2) Outlet temperature of fluid (°C)

3) Temperature of storage tank (°C)

4) Global solar radiation using pyranometer (1) (W/m^2)

5) Diffused solar radiation using Pyranometer (2) (W/m^2)

6) Ambient Temperature (°C)

7) Wind velocity (m/s)

8) Relative Humidity (% RH)

The data is collected by varying thermic fluid flow rate (LPM) from 0.5 LPM to 1 LPM and inlet temperature of fluid from Ambient to 120°C. Inlet temperature of the fluid is controlled by the heater fixed near the inlet of the receiver tube. The data is collected when sky is clear and during the time form 10:30 am to 02:00 pm, so that intensity of solar radiation does not change much. Table 1 shows data collected as an example.

Table. 1. Data Collection (Continuous fluid flow)

Data Set (1) A
Fluid : Hytherm 600
Mass Flow Rate : 0.5LPM
Wind Speed : 0.5 – 1.5 m/s
Inlet Temperature : 35 to 45°C
Date : 08/12/2012

ime	Ta	IbRb	Ti	To	Qu
	°C	W/m²	°C	°C	Joules
10:30	28.6	442	35.4	62.5	28522.8
10:40	28.7	444	38.4	66.4	29470.0
10:50	28.9	474	38.9	68.6	31259.3
11:00	29.0	515	39.3	71.2	33574.8
11:10	29.2	553	39.7	72.3	34311.5
11:20	29.5	574	40.0	73.2	34943.0
11:30	29.7	547	40.1	74.0	35679.8
11:40	30.2	556	40.4	75.4	36837.5
11:50	29.7	609	40.3	77.0	38626.8
12:00	29.3	586	40.0	76.2	38100.5
12:10	29.8	575	40.0	75.8	37679.5
12:20	29.8	581	40.2	77.2	38942.5
12:30	30.2	567	40.7	78.9	40205.5
12:40	30.8	541	41.1	80.8	41784.3
12:50	30.9	520	41.3	82.5	43363.0
13:00	30.9	473	41.5	80.0	40521.3
13:10	31.1	436	41.5	73.1	33259.0
13:20	31.2	408	41.8	74.3	34206.3
13:30	31.0	380	41.9	74.5	34311.5
13:40	31.3	327	42.0	72.6	32206.5
13:50	30.9	322	41.9	69.1	28628.0
14:00	31.4	295	42.0	67.7	27049.3
14:10	31.2	287	42.3	67.1	26102.0
14:20	31.2	277	42.3	65.7	24628.5
14:30	31.4	290	42.5	64.7	23365.5

Data Set (1) B
Fluid : Hytherm 600
Mass Flow Rate : 0.5LPM
Wind Speed : 0.6 – 1.4 m/s
Inlet Temperature : 40 to 55°C
Date : 09/12/2012

Time	Ta	IbRb	Ti	To	Qu
	°C	W/m²	°C	°C	Joules
10:30	28.4	512	42.8	69.8	28687.5
10:40	28.5	423	43.2	71.0	29537.5
10:50	28.6	549	43.8	72.0	29962.5
11:00	28.8	590	44.0	74.5	32406.3
11:10	28.9	633	43.9	75.1	33150.0
11:20	29.2	651	43.8	75.2	33362.5
11:30	29.6	563	44.3	74.5	32087.5
11:40	29.7	622	45.0	76.6	33575.0
11:50	30.2	708	45.6	80.0	36550.0
12:00	30.2	646	46.0	79.4	35487.5
12:10	30.2	633	46.3	80.1	35912.5
12:20	31.2	656	47.0	81.2	36337.5
12:30	31.2	631	47.2	81.3	36231.3
12:40	30.3	604	46.9	83.7	39100.0
12:50	30.7	590	47.3	84.9	39950.0
13:00	30.9	501	48.1	84.6	38781.3
13:10	31.3	467	48.4	81.2	34850.0
13:20	31.4	492	48.9	82.4	35593.8
13:30	31.6	375	49.5	79.1	31450.0
13:40	31.7	370	49.8	78.6	30600.0
13:50	31.6	402	50.3	77.5	28900.0
14:00	31.6	400	50.6	76.7	27731.3
14:10	31.7	374	50.9	75.5	26137.5
14:20	31.8	366	51.5	74.5	24437.5
14:30	31.8	382	51.9	72.5	21887.5

The useful heat collected by the thermic fluid during time interval Δt is given by

$$Q_u = \dot{m} C_p (T_o - T_i) \Delta t$$

Where,

Q_u is the useful heat gain during time interval $\Delta t\ (Joules)$

C_p is the specific heat of the fluid (J/Kg °C)

T_i is the inlet temperature of the fluid (°C)

T_o is the outlet temperature of the fluid (°C)

2.1.2. Instantanious Efficiency of Flsc System

The instantaneous efficiency of the FLSC system is evaluated using following equation [3]

$$\eta = \frac{Q_u}{A_p I_b}$$

Where,

Q_u is the heat gain from the collector during time interval of 10 minutes

A_p is the aperture area of the Fresnel lens and

I_b is the intensity of solar beam radiation received during time interval of 10 minutes

The incident solar radiation consists of diffuse radiation and beam radiation [3]. However solar concentrator's uses only beam radiation to focus on the receiver [3]. We have measured Global radiation and diffused radiation falling on the horizontal surface. Therefore intensity of solar beam radiation falling on the horizontal surface is given by,

$$I_{bh} = I_G - I_D$$

Where,

I_{bh} is the intensity of beam radiation falling on horizontal surface,

I_G is the intensity of global radiation on horizontal plane,

I_D is the intensity of diffused radiation on horizontal plane.

The intensity of beam radiation falling on the aperture of the Fresnel lens is given by

$$I_b = R_b I_{bh}$$

Where,

I_b is the intensity of beam radiation falling on the aperture of FLSC

R_b is the tilt factor

2.2. Steam Generation Test Method

Solar concentrators are mostly used for generating steam or heating thermic fluid [11]. Low pressure steam is required in many process industries such as food processing, dairy, chemical and pharmaceutical industries. Hence we decided to test the performance of the FLSC system for steam generation.

The system used for generating the steam is basically closed loop system based on thermosyphon principle. The thermic fluid system was suitably modified to generate the steam. The schematic view of the system is shown in the Fig.2. The system consists of;

1) Fresnel lens solar concentrator system with lens, receiver and tracking system

2) Small boiler place above the FLSC to generate steam

3) Thermosyphon system to circulate water through receiver.

4) Two temperature sensor T_i and T_o fixed at inlet and outlet of the receiver

5) Three temperature sensors fixed to boiler (T_1, T_2, T_3) to measure temperature of water in the boiler (T_3), Temperature of Steam (T_2) and temperature of outgoing steam (T_1).

6) Two pressure gauges to measure the pressure of steam inside the boiler and when steam is released.

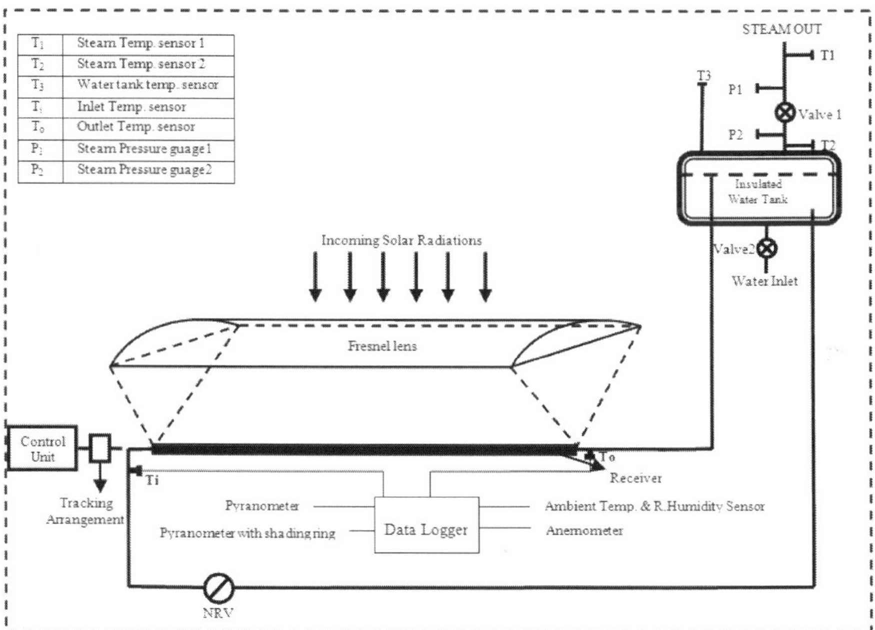

Figure 2. FLSC System Experimental Set-up (Steam Generation)

When solar radiation is concentrated on the receiver, the temperature of the water in the receiver increases and heated water is pushed to the boiler. The equal amount of water from the boiler then pushed to the receiver, form inlet side thus thermosyphon generated a convective flow which circulates water through the receiver. The continuous flow thus generated transfers heat received by receiver to the boiler and temperature of water in the boiler increases. Finally the steam is generated in the boiler. The procedure to measure the performance of the FLSC system is given below.

Initially cold water is filled in the boiler such that about one third volume of boiler is empty. On a clear sunny day, at about 10:30am, the Fresnel lens is adjusted so as to focus solar radiation on the receiver tube. Then the tracking system is started to track the sun. As soon as solar radiation is focused on the receiver tube, water in the receiver starts heating. The temperature gradient between receiver and boiler pushes hot water into the boiler and cold water from bottom of the boiler tank is pushed into the receiver. Thus convective flow is set into system by thermosyphon, which keeps water circulating through the receiver. During the circulation of water, heat is transferred to water which raises the temperature of water in receiver and subsequently in the boiler. Once the temperature of water reaches boiling point, steam begins to generate which is accommodated in the empty space of boiler. When pressure of the steam is above 1.5 – 2.0Kg/cm^2, the valve V_1 is slowly released and steam is exhausted through the outlet pipe of the boiler. The valve is adjusted in such a way that while steam is released, the pressure in the boiler remains constant. This steady state is continued for about 1 hour during which following parameters are recorded by the data logger.

1) Inlet Temperature of water (T_i)

2) Outlet Temperature of water (T_o)

3) Water Temperature in a tank (T_3)

4) Steam Temperature in a Tank (T_2)

5) Temperature of steam when released from boiler (T_1)

6) Global and diffused radiation on horizontal surface (I_G, I_D)

7) Ambient Temperature (°C) and relative humidity (%RH)

8) Wind Velocity (m/s)

Table 2 shows the data collected for a particular day as an example.

Table 2. Data collection for steam

Date: 19/03/2013
Maximum Pressure reached : 1.7Kg/cm^2
Amount water filled in the boiler : 3 Kg
Amount of water left after the release of steam : 0.300 kg
Steam produced by the system : 2.7 Kg
Time when data collection started : 11.00 AM
Time when steam released : 12:00 PM

Time	Solar Radiations(W/m^2)			Temperatures (°C)				
	I_G	I_D	I_bR_b	T_1	T_2	T_3	T_i	T_o
11:00	750.0	151.4	658.5	43.0	36.3	31.0	30.9	89.8
11:10	765.3	152.1	674.5	43.3	36.5	42.2	42.4	75.6
11:20	780.6	152.8	690.6	44.2	38.4	61.0	40.9	78.2
11:30	819.4	152.8	733.3	44.0	42.7	75.8	55.5	81.4
11:40	858.3	152.8	776.1	44.3	47.6	83.6	72.4	91.2
11:50	897.2	152.8	818.9	46.7	51.2	95.2	83.8	100.4
12:00	936.1	152.8	861.7	46.9	50.7	106.4	94.9	110.5
12:10	936.8	154.9	860.1	46.3	57.8	115.8	104.7	118.7
12:20	937.5	156.9	858.6	45.5	62.5	123.4	112.7	126.0
12:30	938.2	159.0	857.1	57.5	109.4	126.0	120.6	127.5
12:40	938.9	161.1	855.6	82.6	110.5	123.8	120.7	124.9
12:50	920.8	166.7	829.6	90.6	108.6	121.3	118.1	122.5
13:00	902.8	172.2	803.6	92.1	109.7	122.8	118.4	123.9
13:10	884.7	177.8	777.6	92.2	113.4	125.6	120.6	126.5
13:20	866.7	183.3	751.7	92.1	113.3	125.8	120.5	126.8
13:30	798.6	182.6	677.6	93.2	112.6	125.5	120.1	126.4
13:40	730.6	181.9	603.5	89.5	106.0	118.0	112.3	116.0

2.2.1. Heat Gain and Efficiency of FLSC System for Steam Generation

Thermal performance of FLSC system for steam generation is carried out as follows,

When Fresnel lens focuses solar radiation on the receiver tube, heat is generated by absorbing solar radiation. Fluid in the receiver gets heated and it is pushed to boiler, thus thermosyphon is activated. The temperature of water in the boiler increases due to heat transfer from receiver to boiler. When boiling point is reached steam is generated in the boiler. Therefore useful heat received by the boiler is,

$$Q_{ub} = mass\ of\ water\ in\ the\ boiler X C_p X \Delta T + mass\ of\ boiler X C_{pb} X \Delta T$$
$$+ mass\ of\ water\ converted\ into\ steam X LH$$
$$Q_{ub} = m_w C_{pw}(T_f - T_i) + m_b C_{pb}(T_f - T_i) + x m_o LH$$

Where,

m_w is mass of water filled in the tank (kg)

C_{pw} is specific heat of water (KJ/Kg°K)

T_f is Final Temperature of Water (°C)

T_i is initial Temperature of water (°C)

m_b is mass of boiler (kg)

C_{pb} is specific heat of boiler material (KJ/Kg)

x is dryness factor of steam ~ 1.0

m_o is mass of steam generated (kg)

LH is latent heat of vaporization of water (KJ/kg)

When we release steam at constant temperature and pressure $\Delta t = 0$

Therefore efficiency of the system given by,

$$\eta = \frac{Q_{ub}}{I_{bn} A_p t}$$

Where,

I_{bn} is intensity of beam solar radiation

A_p is Aperture area of the FLSC and

t is time interval i.e. time during which steam at constant pressure is released

3. RESULTS AND DISCUSSION

The performance of the Fresnel Lens Solar Concentrator (FLSC) system depends on several parameters. Some of the factors affecting the performance of FLSC system are,

1) Optical efficiency of FLSC system
2) Heat removal factor of the system
3) Mass flow rate of fluid used for extracting heat from the receiver
4) Overall heat loss factor
5) Inlet temperature of the fluid and
6) Intensity of solar radiation

The effect of above parameters on the performance of FLSC system is studied using the data collected as described in Table 1 and Table 2.

3.1. Effect of Mass Flow Rate of Fluid on Efficiency

After collecting the performance data for FLSC system using continuous fluid flow test method the data has been analyzed to evaluate performance of the system. Using hytherm 600 as thermic fluid for removing heat, Optical efficiencies at different mass flow rates are as below,

For 1st set of experiment, where flow rate is 0.5LPM

$\eta = 0.51 - 1.37\ \Delta T/I_bR_b$,

Where $\Delta T = (T_m\text{-}T_a)$ & $T_m = (T_i+T_o)/2$

Hence, Optical efficiency = $F_R(\tau\alpha)_e = 0.51$

Heat Loss coefficient = $F_RU_L = 1.37$

For 2nd set of experiment, where flow rate is 0.7LPM

$\eta = 0.56 - 1.44\ \Delta T/I_bR_b$,

Where $\Delta T = (T_m\text{-}T_a)$ & $T_m = (T_i+T_o)/2$

Hence, Optical efficiency = $F_R(\tau\alpha)_e = 0.56$

Heat Loss coefficient = $F_RU_L = 1.44$

For 3rd set of experiment, where flow rate is 1 LPM

$\eta = 0.60 - 1.63\ \Delta T/I_bR_b$,

Where $\Delta T = (T_m\text{-}T_a)$ & $T_m = (T_i+T_o)/2$

Hence, Optical efficiency = $F_R(\tau\alpha)_e = 0.60$

Heat Loss coefficient = $F_R U_L = 1.63$

An increase in the mass flow rate of the fluid increases the value of the heat transfer coefficient [3]. Due to this, the collector efficiency factor $F_R(\tau\alpha)$ and heat loss factor increases. Hence it was observed that when mass flow rate increases both optical efficiency and heat loss factor ($F_R U_L$) increases.

3.2. Effect of Intensity of Solar Radiation on Efficiency

The effect of Intensity of Solar Radiation on efficiency of FLSC system is shown in the Fig.3, Fig.4 and Fig.5. The variation of efficiency with $(T_m - T_a)/I_b R_b$ has been plotted for three different levels of solar radiation (450 W/m^2, 550 W/m^2 and 675 W/m^2). It was observed that the efficiency of FLSC system is strongly influenced by the intensity of solar radiations. The optical efficiencies at three different Intensity levels are as below,

Sr. No.	Solar radiations (W/m^2)	Optical efficiency
1	450	0.55
2	550	0.46
3	675	0.42

3.3. Effect of Inlet Temperature on Efficiency

The inlet temperature of the fluid affects the efficiency of the FLSC system. As inlet fluid temperature increases the receiver temperature also increases due to the heat gain from the solar radiation. When temperature increases the convective and radiative heat losses from the absorber tube also increase [3]. Due to the increase in the heat losses the overall heat gain decreases and hence efficiency also decreases. The data was collected at different inlet temperature with different mass flow rates which is summarized in the below Table 3. It was observed that as inlet temperature increases the optical efficiency increases. The results are shown in Fig.6.

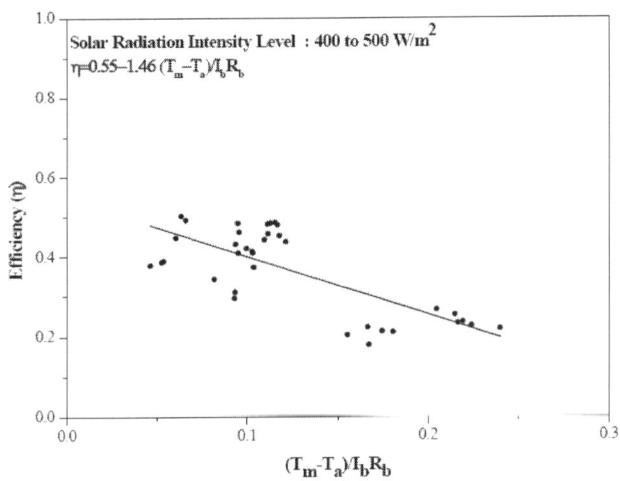

Figure 3. Variation of Efficiency with (Tm-Ta)/IbRb for IbRb = 400-500

3.4. Steam Generation Using FLSC System

The demand for process heat particularly low pressure steam is increasing exponentially. Solar energy can be utilized for generating medium temperature (100°C -150°C) low pressure (T<200°C) steam. By using solar concentrators, low pressure steam can be generated by using low cost technology. Fresnel lens solar concentrators are very useful to generate low pressure steam for industrial and cooking applications. The study of performance of Fresnel lens solar concentrator and steam generation using Fresnel lens solar concentrator is presented here. The overall efficiency of FLSC System is found to be 51% and can generate process steam required for the industry. Table 4 gives the summary of the steam of output of steam when steam is released at constant pressure. The maximum pressure measured to be of 1.8 Kg/cm^2. It was also observed that the steam generation depends on the intensity of solar radiation. The higher the intensity, steam generation will be more. It also revealed that at higher temperature more steam will be generated.

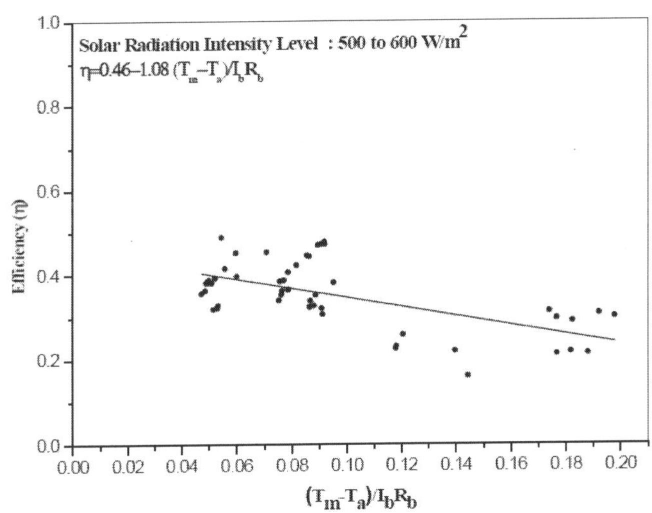

Figure 4. Variation of Efficiency with (Tm-Ta)/IbRb for IbRb = 500-600

4. CONCLUSIONS

Refracting type solar concentrator has been designed and developed which can be used for generating heat in the temperature range of 100°C to 200°C. This Fresnel Lens Solar Concentrator system is useful to generate low pressure steam for various applications like industrial process heat and steam cooking. One of the advantages of this refracting concentrator system over reflecting type solar concentrator system is the durability of the system.

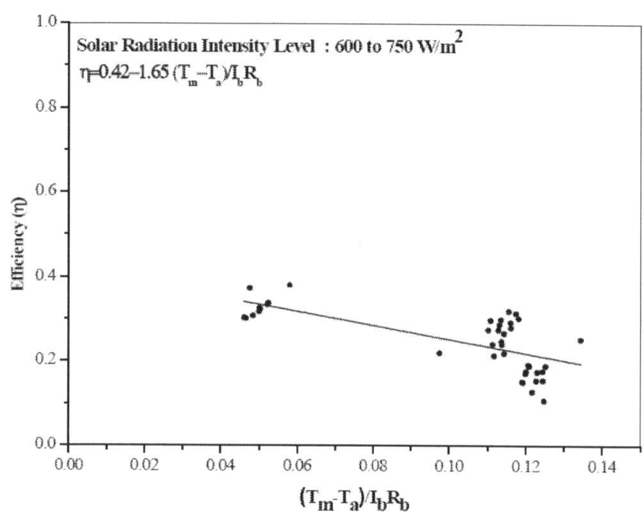

Figure 5. Variation of Efficiency with (Tm-Ta)/IbRb for IbRb = 600-750

It has been found that the Fresnel Lens has life of more than five years which is important aspect for cost effective application, where as for reflecting type solar concentrator, the mirrors need to be changed for every two years.

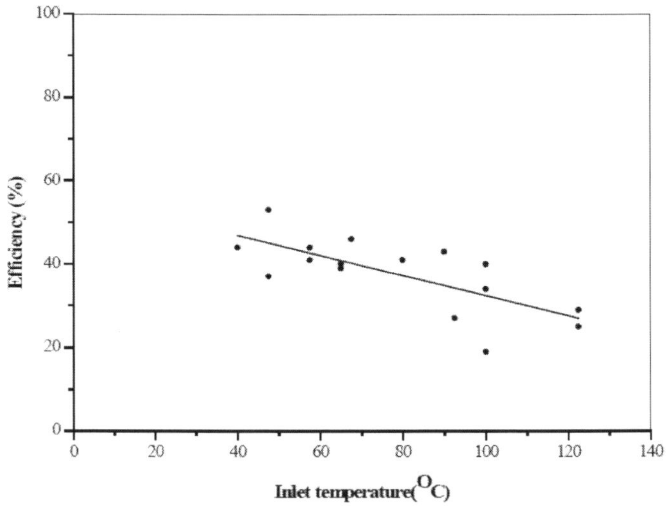

Figure 6. Effect of Inlet Temperature on efficiency

Table 3. Effect of inlet temperature on efficiency

Date	Set No.	Flow rate LPM	Inlet Temperature (°C)	Avg. Efficiency (%)
08/12/2012	1a	0.5	40	44
09/12/2012	1b		50	37
10/12/2012	1c		60	41
13/12/2012	1d		65	39
15/12/2012	1e		90	27
16/12/2012	1f		100	19
17/12/2012	1g		120	25
21/12/2012	2a	0.7	60	44
22/12/2012	2b		70	46
23/12/2012	2c		80	41
25/12/2012	2d		100	34
26/12/2012	2e		120	29
28/12/2012	3a	1	50	53
29/12/2012	3b		65	40
03/01/2013	3c		90	43
04/01/2013	3d		100	40

Table 4. Steam generation Efficiency of FLSC system

DATE	TIME INTERVAL	IbRb	Ta	STEAM	PRESSURE	H. GAIN	S.R.In.	Efficiency
		W/m²	°C	Kg	Kg/cm²	Joules	W.sec	
09/02/2013	12:30 - 13:10	734.6	20.3	0.70	1.3	1851092.0	3755479.7	0.49
18/02/2013	12:05 - 12:30	715.0	22.2	0.90	1.4	2134514.0	4873440.0	0.44
20/02/2013	12:20 - 12:45	555.0	23.1	0.45	1.6	1245079.5	2837057.8	0.44
19/03/2013	11:55 - 13:30	787.5	29.4	2.70	1.7	6135853.0	13419000.0	0.46
20/03/2013	12:20 - 13:00	795.6	31.2	1.40	1.6	3428750.0	6778512.0	0.51
21/03/2013	12:40 - 13:00	877.5	31.3	1.35	1.8	3025609.0	5981040.0	0.51

ACKNOWLEDGEMENTS

The authors thankful to Prof. S.V. Ghaisas, Director, School of Energy Studies, University of Pune for providing laboratory facilities for this work. The authors thankful to Prof. S.R. Jadkar, Department of Physics, University of Pune, Dr. A.M. Pathak, Thermax Ltd. Pune, Mr. S.D. Gunjal, Mr. Prasad Chaudhary and Mr. Onkar Shinde for their continuous help and support during this research work.

REFERENCES

1. Development and study of Fresnel lens based solar concentrator, A.M.Pathak, PhD Thesis, University of Pune, Pune (2009).

2. An experimental investigation of a natural circulation heat pipe system applied to a parabolic trough solar collector steam generation system, Liang Zhang, Wujun Wang, Zitao Yu, Liwu Fan, Yacai Hu, Yu Ni, Jianren Fan, Kefa Cen, Solar Energy, Volume 86, Issue 3, March 2012, Pages 911-919.

3. "Solar Engineering of Thermal Processes", Third Edition (Year 2006) by John A. Duffie and William A. Beckman.

4. Procedure for thermal performance evaluation of steam generating point-focus solar concentrators, Vishal R. Sardeshpande, Ajay G. Chandak , Indu R. Pillai Solar Energy 85 (2011) 1390–1398.

5. "Solar Heat for Industrial Processes, Medium Temperature Collectors", Report of International Energy Agency IEA), Task 33/IV, Publisher: AEE INTEC, Feldgasse 19, A-8200 Gleisdorf, Austria, May, 2005.

6. "Development and Optimization of a Parabolic Trough Collector System for Generation of Industrial Process Heat", www. fabrikderzukunft.at.

7. "Solar Concentrator for Industrial Process Heat,' Update", A newsletter of Industrial Research and Consultancy Centre, IIT Bombay, Issue-1, 2006.

8. Gadhia Deepak, "Successful use of Solar Concentrators for various applications from cooking to process needs for rural development: Indian Experience", www.gadhiasolar.net.

9. Design of a Non imaging Fresnel lens for solar concentrators, Ralf Leutz, Akio Suzuki, Atsishi Akisawa and Takao Kashiwagi Solar Energy Vol. 65, No.6, pp.379–387, (1999).

10. "PTC 1000 - Modular parabolic trough collector", estec 2005, the 2nd European Solar Thermal Energy Conference, 21-22 June 2005 Freiburg.

Chapter 10

Link Quality-Based Transmission Power Adaptation for Reduction of Energy Consumption and Interference

Jinglong Zhou, Martin Jacobsson, and Ignas Niemegeers

Faculty of Electrical Engineering, Mathematics and Computer Science, Delft University of Technology, P.O. Box 5031, 2628 CD Delft, The Netherlands

ABSTRACT

Today, many wireless devices are mobile and battery powered. Based on the fact that battery capacity is still limited, energy saving is an important issue in wireless communication. Meanwhile, the number of wireless devices continues to increase and this creates interference problems between wireless devices. In this paper, we look at transmission power control and propose a mechanism that tries to achieve minimum energy consumption or emission under any circumstance. Lower transmission power levels may result in more retransmissions, but in total, energy consumption or emission still can be reduced in many scenarios. To evaluate the performance of our mechanism, we used real wireless channels in an indoor environment to carry out measurements. The measurement results indicate that a significant amount of energy consumption or emission reduction can be achieved for the transmitter in most scenarios compared to using a fixed transmission power level for all packets.

1. INTRODUCTION

Plenty of wireless devices use battery-based power, but the battery technology does not keep up. To increase device service duration, saving power is crucial. Power saving in communication can be achieved by different methods at different communication layers. Power-aware routing selects routes that together consume less energy or use devices that have more energy [1]. In the MAC layer, the receiver can turn off the receiver function periodically to save energy [2]. Another way of saving energy is to adapt the transmission power for the transmission of packets. Power transmission adaptation can achieve two benefits: save energy and reduce interference. Interference is becoming an increasing problem due to the enormously growing number of wireless devices. One way to alleviate this problem is to reduce the emitted transmission power.

The motivation for transmission power adaptation for energy saving and interference reduction stems from the fact that many of the current wireless communication systems (e.g., IEEE 802.11 and IEEE 802.15.4) usually use a fixed default transmission power level for all transmissions. However, when two nodes are very close to each other, the default power level is much higher than required to successfully deliver all packets. This both wastes energy and creates unnecessary interference. A lower transmission power level may require a larger number of retransmissions, but overall less energy will be emitted or consumed for each transmission and in total, there may be less waste. Therefore, a trade-off is possible between the number of retransmissions and energy consumption for each packet delivery. This trade-off requires the knowledge of the packet delivery ratio (PDR) for each transmission power level. We call this the *PDR-table*. The PDR-table differs between different links and different environments. To always select the transmission power level that consumes the least energy or have lowest energy emission, a self-adaptive transmission power adaptation mechanism is required that accurately observes the PDRs. In this work, we focus on IEEE 802.11 and IEEE 802.15.4 as our experiment technology. However, our methods can be used in other radio technologies as well.

Energy consumption for IEEE 802.11 is not so crucial as for IEEE 802.15.4, since IEEE 802.11 is normally used with larger devices, such as laptops, PDAs, and mobile phones, which can be recharged easily. However, minimizing energy emission is still important because of the interference. For IEEE 802.15.4, energy consumption is critical due to its use in wireless sensor networks. Therefore, we mainly discuss interference reduction for IEEE 802.11 and energy saving for IEEE 802.15.4.

In this paper, we propose a power transmission control mechanism that is based on gathering PDRs for every transmission power level (the PDR-table). It consists of two phases: *initialization* and *updating*. It can be used both as an interference reducing mechanism and an energy saving mechanism depending on the energy model. We propose five different methods for the initialization phase. In the updating phase, we use an exponential weighted moving average (EWMA) method to update the PDR for each transmission power level and use the result to select the optimal level. To the best of our knowledge, we are the first to select the transmission power that achieves the minimum energy consumption or emission for delivering a certain amount of information based on link PDR-tables. We explore the maximum potential reduction of energy emission and consumption by an investigation of all relevant parameter combinations in our mechanism. The proposed mechanism is evaluated based on measurement data and the results indicate that significant savings can be achieved in many scenarios compared to always using the default transmission power level. We also compare our

PDR-based mechanism with one that uses signal strength. Also there, the results indicate a significant improvement.

The rest of this paper is organized as follow: Section 2 introduces related work and Section 3 presents our measurement results and shows the potential reduction of energy consumption and emission. Our PDR-based transmission power adaptation mechanism is introduced in Section 4. In Section 5, our experimental system is described and in Section 6, the measurement results are presented. The paper is concluded in Section 7.

2. RELATED WORK

Transmission power control requires good knowledge of the correlation between link quality and transmission power levels. This correlation has been studied before via measurement activities. In [3, 4], the correlation of transmit power level and packet delivery probability was analyzed in different indoor scenarios. Based on their observations, small adaptations in the power level do not change the packet delivery ratio in any measurable way. Some work also discussed combinations of power and rate adaptation to achieve good performance. In [5], it was proposed to select data rate and transmission power based on link quality. The method was applied in an indoor environment and achieved higher throughput than the traditional mechanism. However, energy consumption was not calculated.

Most previous work on applying transmission power adaptation schemes was more focused on reducing interference, maintain connectivity, and topology control, such as [6, 7, 8, 9]. Paper [10] discusses the use of transmission power control to select reliable links and disable unreliable links via a blacklisting method in order to improve the system performance. Paper [11] discusses the use of transmission power control to reduce interference and simulation results reveal that throughput can be increased by adapting the transmission power in an ad hoc network. This shows the benefit of reducing energy emission. However, the aim of these papers were to maintain the link quality at a certain level, control the topology, and increase throughput by using transmission power adaptation. Energy was not their main focus and the selected transmission power level does not always result in the minimum energy consumption or emission level.

A few papers address energy saving explicitly. The authors of [12] proposed to use a RTS-CTS handshake in the highest power level to discover the channel quality and then use the lowest possible power level for the data packet. Simulation results show that the proposed power mechanisms can achieve energy savings without degrading the throughput. However, in their proposal, a separate channel is used for controlling, which means that adaptations to the IEEE 802.11 standard are necessary. Meanwhile, a theoretical model does not reflect the real channel situation accurately. In [13], a loop-based mechanism is used to adapt the transmission power level to achieve the minimum required power level for message delivering. Simulation results show that energy can be saved and throughput can be increased. However, this work also assumes that a RTS-CTS handshake is used. Moreover, a mechanism that adapts the transmission power level one level at the time will be too slow for fast channel variances. It may take several periods for the system to choose the appropriate power level.

In [14], the authors propose a power saving algorithm that adjusts the transmission power and extends the network lifetime. Again, only simulations are used to validate the proposed protocol. Paper [15] is the most similar work to ours; transmission power adaptation was used for power saving in different scenarios. However, the optimal transmission power level is set by the received signal strength. We use PDR information for two reasons. First of all, the mapping between PDR and received signal strength is not straight forward and noise and interference have a large impact on the mapping. Second, different receivers have different sensitivity levels and using received signal strength may require different thresholds for different devices. A PDR-table method is affected by different devices. We compare this mechanism with our mechanism in Section 6.

3. ENERGY EMISSION AND CONSUMPTION MEASUREMENTS

To minimize the energy consumption or emission for successfully delivering a fixed amount of information, such as a certain number of packets, we turn to the expected energy consumption or emission. We calculate the expected total energy consumption or emission for one packet delivery as follows:

$$E = P \cdot N \cdot T, \tag{1}$$

where E is the total energy consumption or emission for successfully delivering one packet (in Joules). P is either the energy emission or the energy consumption (in Watt), N is the expected required number of transmissions to successfully deliver a packet (i.e., $N = 1 / \text{PDR}$), and T is the duration (in seconds) for one packet transmission including headers and preambles. We can see that if we use a single data rate and packet-size, T will be a constant value. E can be calculated for each transmission power level and the result can be used to find the optimal level, that is, the one with the lowest E. Depending on what P value we use, we will optimize for different things. For instance, if we are interested in minimizing energy emission we use the following formula:

$$P = P_{\text{RF}}, \tag{2}$$

where P_{RF} is the energy emission created by the transmission power level. For IEEE 802.11, the transmission power range is from 0 to 15 dBm and for IEEE 802.15.4, it is from −25 to 0 dBm [16]. Our 802.15.4 device has 31 different power levels, but we used only 15 of them, which we calculate in this simplified way: level 3 corresponds to −23 dBm and level 31 corresponds to 0 dBm and then we assume a linear correlation to map the transmission power levels in between to the different energy emission levels in dBm.

For minimizing the energy consumption, we also need to consider the energy consumption of the wireless device circuit, the energy consumption ($P_{E_{TX}}$) of other parts, and the wireless card amplifier energy consumption (P_S) as shown in Figure 1. While P_S is dependent on the transmission power level, $P_{E_{TX}}$ is not.

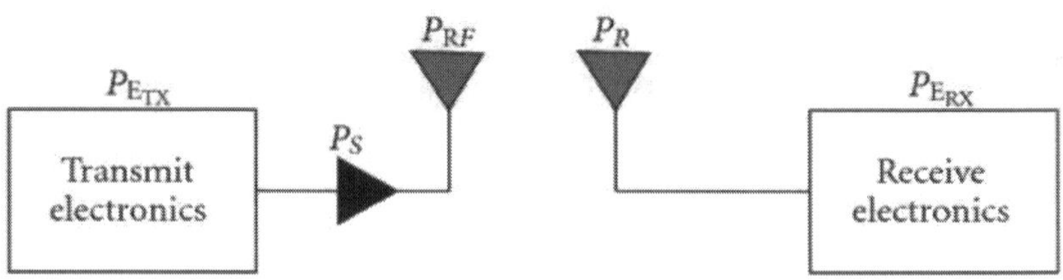

Figure 1. The high level block model of an RF link.

For calculating the total energy consumption, we refer to the results in [17, Figure 5]. Since measuring the PDR introduces a lot of inaccuracies, we do not need a perfect approximation of the energy consumption. Hence, we can simply use the following linear equations for approximating the energy consumption:

$$P = 10 \cdot P_{RF} + 1400; \quad \text{(for IEEE 802.11)}, \tag{3}$$

$$P = 35 \cdot P_{RF} + 30; \quad \text{(for IEEE 802.15.4)} \;. \tag{4}$$

If we only calculate the energy emission to the environment, (1) and (2) are used. If we calculate the total energy consumption of the whole transmitter, (1) and either (3) or (4) are used.

To capture the accurate correlation between transmission power and PDR, a measurement-based method has to be used. For this reason, we carried out measurements in an indoor environment with different radios and configurations. For all experiments, the same number of packets (2000) were sent with 15 different transmission power levels. Two different radio technologies were used, IEEE 802.11 and IEEE 802.15.4. Let us first start with IEEE 802.11. We used UDP with a fixed packet-size of 1500 Bytes including the IP header due to the fact that this packet-size is common in the Internet traffic [18]. We ran some indoor scenarios with different locations, but with a fixed data rate. Then we tried different data rates in the same location. The results are presented in Figure 2. The first group of experiments were done with 2 Mbps data rate in three different scenarios, using different distances between the sender and the receiver. The measurement PDR-table of the three stationary scenarios is plotted in Figure 2(a). The second group of experiments were done with different data rates and are presented in Figure 2(b). All scenarios and the experiment setup details are further described in Section 5.

At the receiver side, we recorded the PDR for each transmission power level. When doing this for our scenarios, we obtained the results in Figures 2(c) and 2(e). We can see that a certain transmit power level achieves the minimum energy emission or consumption and they are different for different links. The minimum energy emission level for each link in Figure 2(c) is 3, 6 and 9 for each link, respectively. For the energy consumption, we use log scale to show the results due to the large differences. We can still see that there is a level which results in the lowest energy consumption for the transmitter, and this level is not the highest power level.

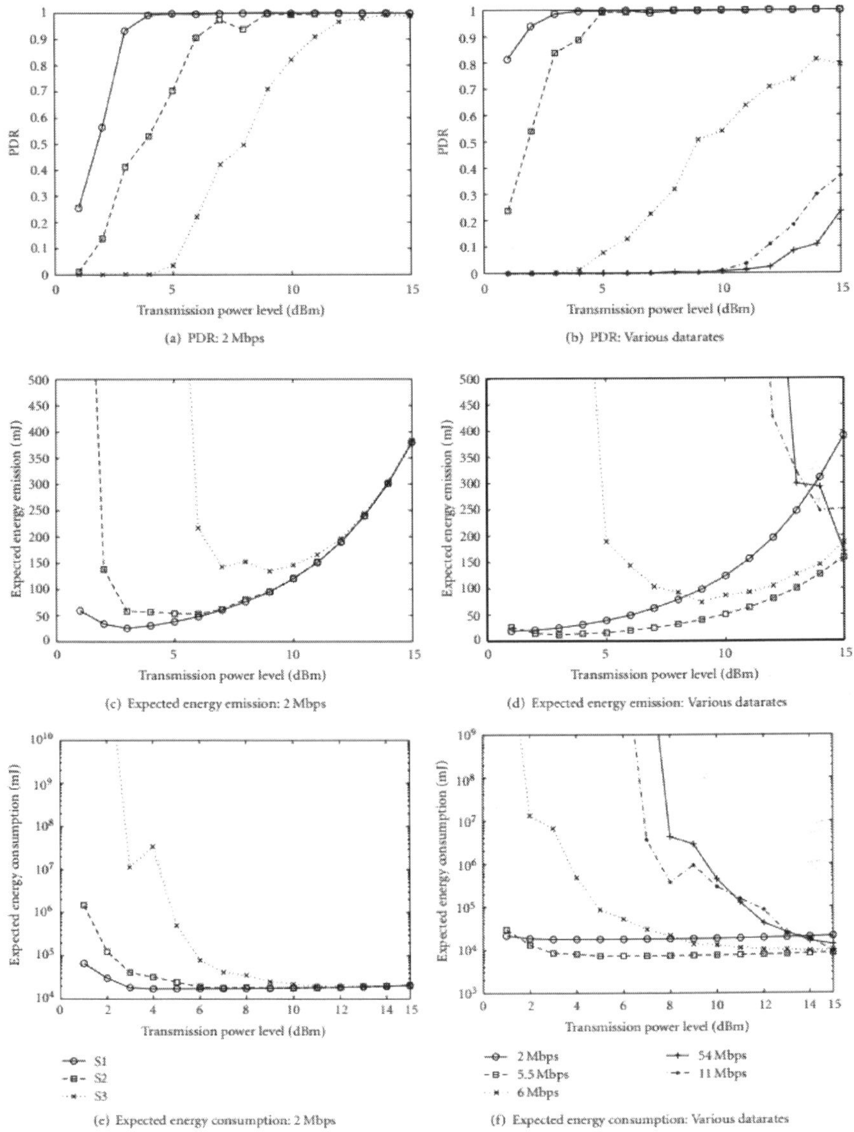

Figure 2. The PDR-table and expected energy emission and consumption for IEEE 802. 11.PDR: 2 MbpsPDR: Various dataratesExpected energy emission: 2 MbpsExpected energy emission: Various dataratesExpected energy consumption: 2 MbpsExpected energy consumption: Various datarates

To show that this phenomenon not only exists for IEEE 802.11 with 2 Mbps data rate, we carried out measurements for many data rates. The power trade-off for IEEE 802.11 with different rates is presented in Figures 2(d) and 2(f). It is interesting to see that for higher data rates, for example, 54 Mbps, the level that results in minimum energy consumption and emission is 15. This is caused by the fact that the link quality is so poor and struggles even with full power.

In Figure 3(a), the PDR-table with different transmission power levels but with a fixed packet-size in IEEE 802.15.4 is presented. We can see that although the power level is different from IEEE 802.11, the results are similar to Figure 2(a). For IEEE 802.15.4, only one data rate is possible, but we can change the packet-size. When we change the packet-size in Figure 3(b), we can see some PDR changes. However, the PDR difference is not very obvious. We also calculated the expected energy emission and consumption for IEEE 802.15.4 and present the results in Figures 3(c) and 3(e). The power trade-off for IEEE 802.15.4 with different packet-sizes is presented in Figures 3(d) and 3(f). The expected energy emission and consumption are calculated and compared with the case where we assume that every link had to deliver the same amount of bytes. We used 100 Byte as assumed payload, which means for a 20 Byte packet payload, one needs to deliver five packets to reach the same information delivery. In the same way, one needs two 50 Byte packets.

Based on the four groups of results shown in Figures 2 and 3, we can see that almost all the links have a PDR from 0 to 1 within a 10 dBm transmission power difference. In almost all situations, the PDR is higher for larger transmission power levels. From Figures 2(c) and 3(c), we can see that given a data rate and packet-size, links with better PDR always requires less energy emission and consumption to deliver the same number of packets. However, if we are also able to change the data rate and packet-size, it is possible to further lower the energy emission and consumption.

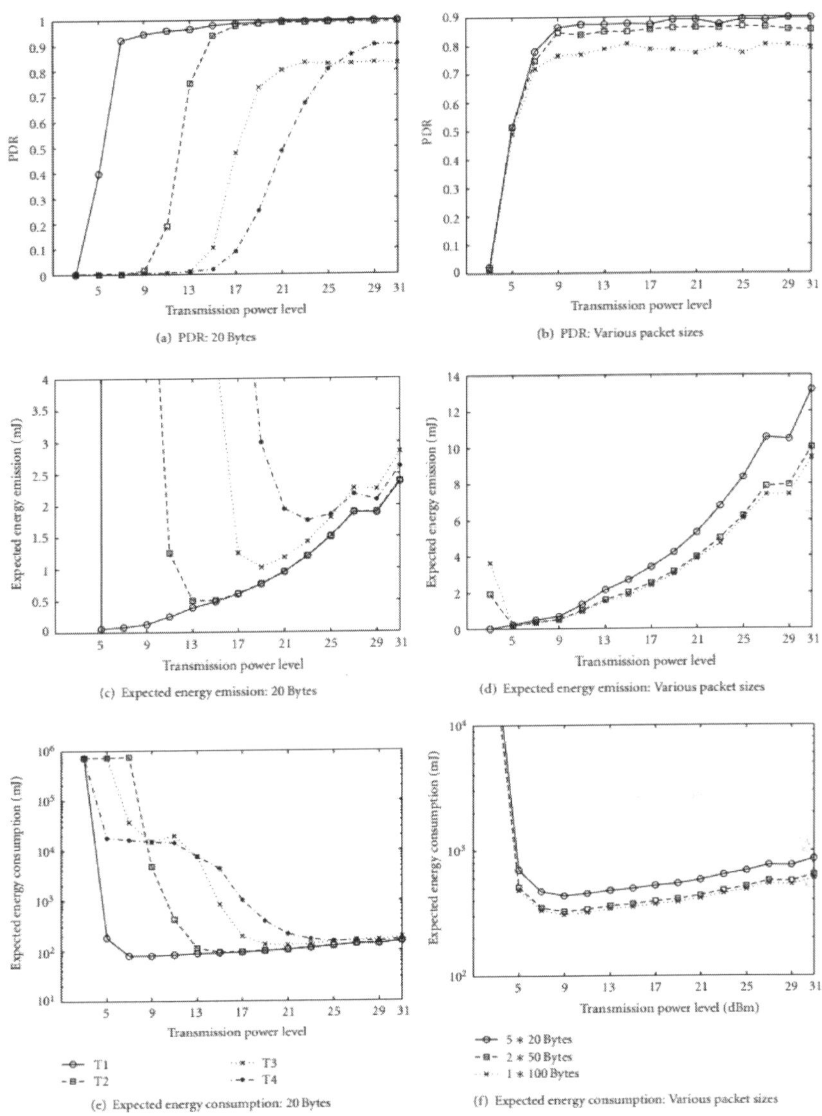

Figure 3. The PDR-table and expected energy emission and consumption for IEEE 802. 15.4.PDR: 20 BytesPDR: Various packet sizesExpected energy emission: 20 BytesExpected energy emission: Various packet sizesExpected energy consumption: 20 BytesExpected energy consumption: Various packet sizes

4. PDR-BASED TRANSMISSION POWER CONTROL

For a certain channel, if the correlation between P, N, and T is known and constant, the best combination can be selected easily. However, the actual channel PDR-table can be quite different from link to link as shown in Figures 2and 3 and this is also indicated in [3]. Therefore, to have an efficient transmission power control, we need a good mechanism of learning this PDR-table in real time. Meanwhile, the PDR-table may change due to several reasons, such as mobility, environmental changes, and interference. Hence, a self-adapting mechanism is required.

For each link, we need to keep a PDR-table that contains all the N values for the different transmission power levels. The PDR-table may contain values for all possible transmission levels or only a subset of them. The P values are not dynamic and can be calculated beforehand for each of the transmission power level based on the chosen energy model. Since (1) will be used for both the energy emission and consumption calculation, we can use the same transmission power control mechanism for both.

We divided the mechanism into two phases; the initialization phase and the updating phase. The initialization phase tries to quickly learn or "guess" the correlation between the transmit power level and the PDR once a new communication link is established. The updating phase keeps on updating this PDR-table and adapts the transmission power during the whole communication period. The initialization phase should be very short compared to the updating phase. Hence, the initialization phase is more useful for small amounts of traffic and the updating phase is more useful for large amounts of traffic. We describe the two phases in detail in the following two sections.

For neither phase, we do not generate any extra packets to probe the PDR-table. Instead, we use the normal data packets to "learn" the channel and select the appropriate transmission power level. If acknowledgments are being used, which is the case for most wireless links, including 802.11 and 802.15.4, the sender can use them to find out about the packet losses. Otherwise, this information needs to be passed back to the sender in another way. The energy emission or consumption calculation for all methods have the same prerequisite, the same amount of information need to be delivered.

4.1. Initialization Phase

In the initialization phase, different methods can be used to learn or "guess" the correlation between PDR and transmission power level and populate the PDR-table. We propose four initialization methods and compare them with the default method that always transmits with maximum transmission power, which we call "Fixed". We introduce all four methods as follow:

(i)Default start. Start using the default power level (15 dBm in 802.11 or 0 dBm in 802.15.4) and then immediately move on to the updating phase. This means only one packet is transmitted and depending on whether it was received or not $N = 0$ or ∞ for the default power level. The remaining N s in the PDR-table are set to ∞ .

(ii)Sampling. Send 10 packets in all transmission levels to probe the channel and then use the obtained measurements to build the initial PDR-table and then move on to the updating phase.

(iii)Historical. Use the last recoded PDR-table (recorded based on the latest communication record between two nodes). The sender sends 10 small packets (40 Bytes) with full transmission power and the receiver reads and sends back the received signal strength. The sender then compares this with the received signal strength recorded last time. The original table is shifted left or right with the difference value based on the signal strength difference and forms the new PDR-table.

(iv)Combined. First collect the received signal strength as in the Historical method. If the signal strength between now and the previous communication are similar (within 2 dBm difference), the Historical method is used. Otherwise, the Sampling method is used.

A better initialization method starts closer and converges faster to the optimal transmit power. In Section 6.1.1, we will compare all these methods with the Fixed method, which sends all packets with default power level during both the initialization and updating phases and hence makes no use of the PDR knowledge.

4.2. Updating Phase

In the updating phase, most packets are transmitted with the transmission power level that minimizes (1). If two levels have the same power consumption, then the higher transmission power level will be used.

The estimated PDR for the other power levels also needs to be updated, since the whole PDR-table is dynamic if the link changes. Therefore, we propose to send a certain percentage of packets using a randomly selected power level other than the current one. In this way, the estimated PDR for all power levels can be updated. Periodically, we calculate the PDR for each level by dividing the number of received packets with the number of sent packets during that period. To have a controllable smooth updating process for all the information, we use an EWMA method as in (5),

$$E_{t+1} = \alpha X_t + (1-\alpha)E_t, \tag{5}$$

where the E_t means the current estimation of PDR for a certain transmission power level in interval t, X_t is the calculation of PDR for this power level in interval t, and the smoothing factor α is used to tune the speed of updating. This is only done for N values that had a transmission in the PDR-table during the interval. We used an interval of 10 packets.

We defined another parameter which controls the probability that a packet will use another level than the selected optimal level. This probability is defined as β . The level to probe is selected uniformly among the other levels in the PDR-table. The performance of the updating phase with different α and β is investigated in Section 6.1.2.

5. EXPERIMENTAL SETUP

All experiments were carried out in a typical indoor office environment. They were done at night when there were very few people walking around. For each scenario, we collected a packet trace and used a post processing approach to compare every method and parameter. In this way, every parameter combination could be compared based on the same actual link in a fair way.

5.1. IEEE 802.11 Test-Bed

For all our IEEE 802.11 experiments, we used two HP laptops (HP7400) equipped with 3Com 108 Mbps 11g XJACK PC wireless cards. Linux 2.6 and the Madwifi driver version 0.9.4 were used. We specially wrote a one-hop communication program, which had a sender and a receiver part. The node running the sender program controlled the transmission power level for each packet transmission. A fixed packet-size (1500 Bytes) was used during all experiments. We used broadcast packets to avoid MAC level retransmissions and the receiver side recorded the number of received packets. In a real system, feedback from the retransmission mechanism can be used instead.

We used channel 7 during the experiments. Long duration observations were done of the noise level for this channel and the value was around −96 dBm with a maximum variance of 2 dBm. Different distances (8, 16, and 20 meters, resp.) were used in the experiments to generate different channel conditions, but always nonline of sight (NLOS). We name these scenarios as S1, S2, and S3. For the experiments with different data rates, we used a distance of 20 m with another NLOS channel. Therefore, we call it S4.

5.2. IEEE 802.15.4 Test-Bed

We used an IEEE 802.15.4 compliant device in the 2.4 GHz ISM band from Moteiv, called Tmote sky that uses the CC2420 wireless chip [16]. During the experiment, the USB was used as power supply. As in IEEE 802.11, we also wrote a one-hop communication program for these devices. We used three different payload sizes. They were 20, 50 and 100 Bytes. IEEE 802.15.4 has a packet header, which consists of 11 Bytes of PHY header and 6 Bytes MAC header. The standard data rate (250 kbps) was used during all experiments. We used only 15 different transmission power levels for the Tmote to be more comparable with our 802.11 experiments. Since there are 31 possible levels, we only used the odd levels between 3 and 31. Based on [16], they correspond to dBm as follows: Level 3 corresponds to −23 dBm, level 31 to 0 dBm and the levels in between are mapped in an almost linear fashion.

All the experiments were done in a channel that did not interfere with any IEEE 802.11 radio. We also did experiments in a channel that was impacted by IEEE 802.11 radio interference and found that the result was not much influenced. We used broadcast packets in the same way as in IEEE 802.11. We recorded the number of received packets and the used transmission power levels.

The IEEE 802.15.4 experiments were done in the same location as for IEEE 802.11, however, different distances were used. All channel were

NLOS and the distances were 12, 14, 16, 18 m, respectively. We call these experiment scenarios T1 to T4. The experiments with different packet-sizes were done with 17 m between the sender and receiver with a NLOS channel.

5.3. Experiment Methodology

For each scenario, we collected a data trace by sending 30000 packets with different power levels during a period of 20 minutes. To be able to compare fairly between different methods and parameters, we used a post processing approach. In this approach, we took the trace and divided it into 200 batches. Each batch contained 150 packets, 10 packets of each power level. For each method and parameter combination, we emulated the process. This was done by assuming that only 10 packets were sent from each batch and it was up to the method to decide which power levels to pick. That is, for each emulation, only a fraction of the trace was used.

For the updating phase, $(1-\beta)$ % of the 10 packets were assumed to be transmitted with the currently selected best power level and β % were assumed to be sent for probing the other power levels. These assumed packets were randomly selected from the trace, based on the power level and the batch it belonged to. From the trace, we checked whether the selected packets were received or not and used this information in the method. An important issue is that, due to the limited number of packets on each nonbest transmission power level (e.g., $10 \cdot 10$ % for each interval is only 1 packet), the PDR for each transmission power level is only updated when there is a packet transmission in this interval. Since this random selection introduces variance, we repeated this process 300 times and calculated the mean and 95% confidence interval.

Parts of the packets are sent in the initialization phase and parts are in the updating phase. Each transmission was done with a certain transmission power level and took a certain duration. Therefore, the total energy emission or consumption was the sum of all energy emitted or consumed for all the transmissions. We processed the data using this method several times and due to some random factors in the processing, the total energy emission from each processing are hardly exactly the same. However, they are quite similar and the confidence intervals are very small, so we did not plot them and only plotted the average expected energy emission for a certain method and parameter combination. We did the same processing for the updating phase as well.

Unfortunately our IEEE 802.11 card did not support fast power variation. Based on measurements, we could conclude that it took our card about 1 second to change from the highest to the lowest transmission power level. Hence, we divided the time into intervals, each of 8 seconds long. In each interval, we first transmitted 200 packets with one transmission power level and then paused for two seconds. Right after the pause, we modified the power level to the next level and waited two seconds. The power level was changed in a round robin fashion between all 15 levels. For IEEE 802.15.4, we changed the power level per packet, which caused no problems.

6. PERFORMANCE EVALUATION

In this section, we evaluate the performance of our PDR-based mechanism. The energy emission and energy consumption are discussed in the following two sections, starting with the energy emission. In Section 6.3, we look at strategies to optimize both.

6.1. Energy Emission Reduction

First, we present the emission reduction results for both the initialization and updating phases.

6.1.1. Initialization Phase

The target of the initialization phase is to quickly populate the PDR-table and select a good transmission power level to start with and then enter the updating phase as explained in Section 4.1. In this comparison, a fixed α value of 0.2 and a fixed percent of probing packets of $\beta = 10$ % were used in the updating phase. We tried different α and βvalues in Section 6.1.2. For the Historical method, we used the PDR-table learned from the same location one day earlier. In Figure 4, we present an example of how each initialization phase selects the best transmission power level in each batch for IEEE 802.11. We can see that all methods, except Fixed, converge to the best transmit power level (around 2 dBm) after no more than 50 batches (corresponding to 500 s or 500 transmitted packets).

We calculated the total expected energy emission for the first 60 batches of each method and present the results in Figure 5. The number of 60 batches is selected due to the reason that after this time, all the methods definitely go to the updating phase. The expected energy emission means the required energy needed to be generated to the environment to deliver a certain amount of information, that is, to successfully transmit all 2000 packets. We can see that all our proposed initialization methods can reduce the energy emission compared to the Fixed method. The Historical and Sampling methods can further reduce the energy emission compared to Default start. The Combined method achieved the best performance, which indicates that using an accurate PDR-table is essential for a good initialization phase.

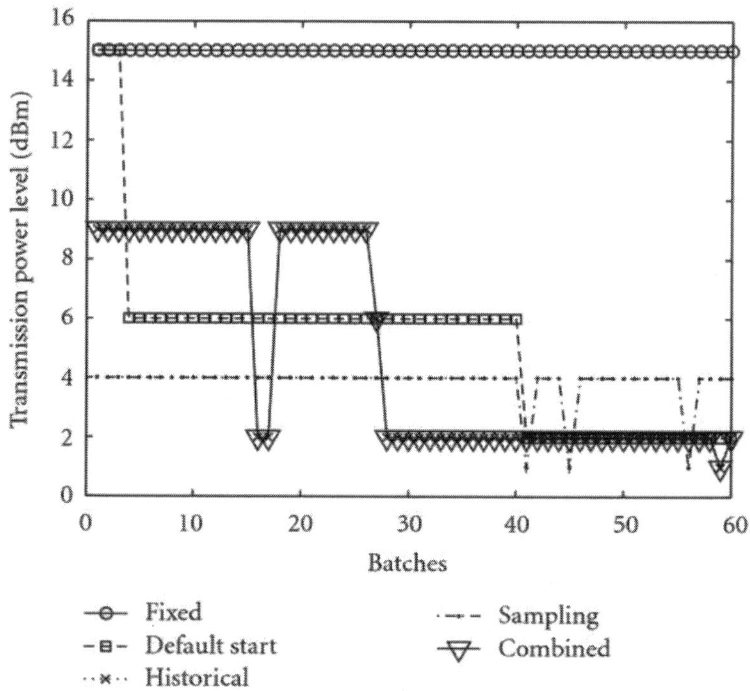

Figure 4. The selected best power level in each time interval by different methods in scenario 1 (IEEE 802. **11).**

6.1.2. Updating Phase

(i) IEEE 802.11

For the updating phase, we need to find the optimal α to use in (5). To have a fair comparison of all different αvalues, we fixed all the other parameters. The percentage of probing packets, β , was set to 10% and we used the Default start method. For each α value, we calculated the average expected energy emission of 300 experiments and show the result in Figure 6(a) based on all 200 batches from the trace. We can see that when $\alpha > 0$, the energy emission decreases compared to when no updating is done ($\alpha = 0$, always using 15 dBm) and that different links have different optimal α. We can also see that when $\alpha > 0.2$, no major improvements can be seen. Since a smaller α is better for mobile scenarios, we propose to use $\alpha = 0.2$.

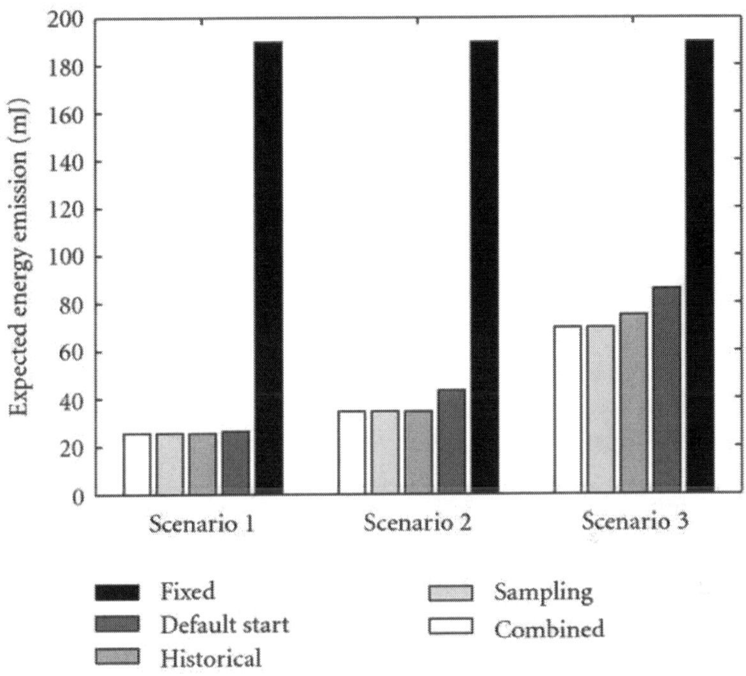

Figure 5. The initialization phase performance comparison (IEEE 802. 11).

Another parameter to investigate is β. Figure 6(b) shows the results of using $\alpha = 0.5$ and different amounts of probing packets. We can see that for each scenario, the optimal β values for each link are all between 5 to 10%, which suggests that we should not send too many packets to probe other transmission power levels. However, the optimal β is different for each link. The general rule is that, when the link is worse (PDR is lower for most power levels), the optimal β is larger, which suggest that for lossy links, more probing should be done. However, a value of 10% performs well enough for all scenarios.

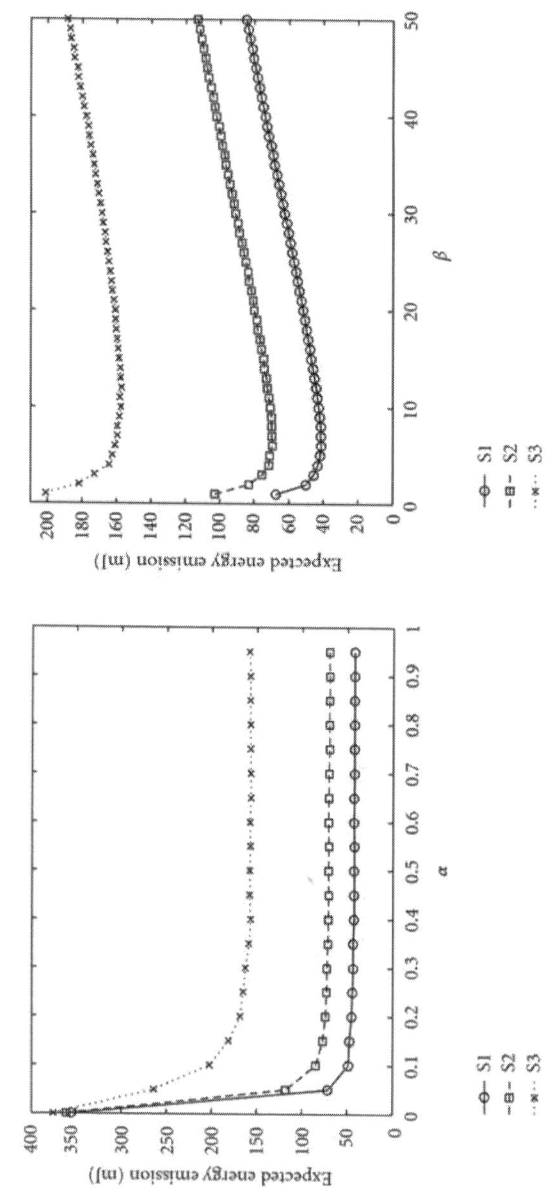

Figure 6. The best α **and** β **for IEEE 802.** 11, 2 Mbps.best α best β

Using $\alpha = 0.2$ and $\beta = 10$ %, we made a general comparison in Table 1 between the PDR-based method and the Fixed method of always using 15 dBm. Default start was used in the initialization phase. We can see that for each scenario, the energy emission is much less than for the Fixed method.

Table 1. Quantitative comparison of expected energy emission for the updating phase: IEEE 802.11.

Scenario	S1	S2	S3
Fixed (mJ)	379.44	379.44	379.44
PDR-based (mJ)	41.87	67.78	161.42
Reduction	−89%	−84%	−57%

(ii) IEEE 802.15.4

We used the same processing code to process the results for IEEE 802.15.4, but with the traces from scenario T1 to T4. To have a fair comparison of all different α values, we fixed β at 10%. We used the maximum transmission power level (31) to start. Based on Figure 7(a), we can see that we got similar results as in Figure 6(a). When α is larger than 0.1, the expected energy consumption is much smaller than the expected energy consumption when αequal to 0. There is not much difference when α is larger than 0.1.

We further processed the measurement results with the assumption that α is equal to 0.5 and we compared the expected energy emission with different β values, from 1 to 50. The results are shown in Figure 7(b). The optimal βvalue for $\alpha = 0.5$ is around 5% and more probes will result in more energy emission.

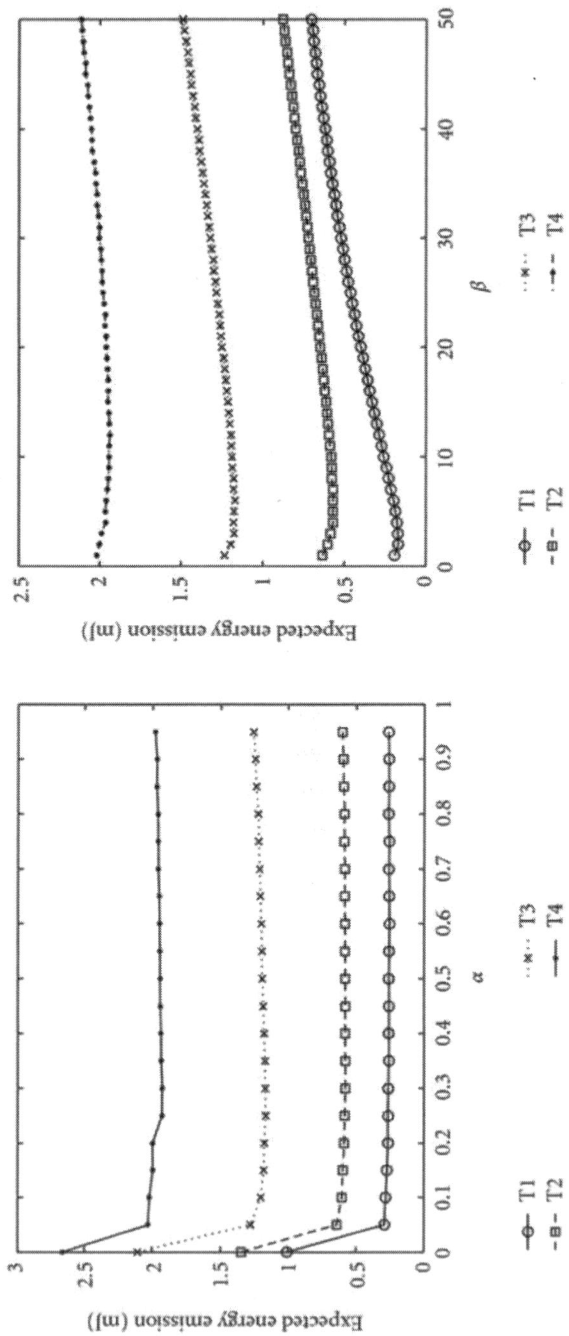

Figure 7. The best α and β for IEEE 802. 15.4, 20 Bytes.best α best β

To have a better comparison between different α and β in each scenario, we calculated all the combinations for α values from 0 to 1 in steps of 0.05 and β values from 1 to 50 in steps of 1.0 . In Figure 8, we use a 3D graph to show the expected energy emission for all combinations. A common trend is that when $\alpha = 0$, which means no update at all and always use the highest transmission power level, the energy emission is much larger compared to when $\alpha > 0$. In Figure 8(a), we can see that it is very obvious that larger β values will result in more energy emission. This is because the optimal transmission power level is 5 and higher power levels will cost more energy for each transmission. Most power levels are not worth to be probed, therefore, a larger β results in more energy waste. When the channel becomes worse, the expected energy emission with different β is less, which is most obvious in Figure 8(d). Another interesting result is that there are more fluctuations when α or β increase in scenarios with worse channels, which can be seen in Figure 8(d).

Similar to Table 1, we calculated the total energy emission for each scenario with $\alpha = 0.2$ and $\beta = 10$ % and present the results in Table 2. The best α and β values are also included in the table. We can see that the PDR-based method only emits about 20% to 53% percent of the energy compared to the Fixed method. We also present the values based on the optimal α and β selection from Figure 8. We can see that in most cases, we are very close to the optimum simply by using $\alpha = 0.2$ and $\beta = 10$ %, which means we can use this combination for almost all the scenarios.

6.1.3. Impacting Factors for the Updating Phase

The results presented in Section 6.1.2 are all with the same configuration for different scenarios. To make it more practical to the real world, we further evaluated the performance of the updating phase by studying factors that may influence the performance, such as transmission data rate, packet-size, mobility, and time dependence.

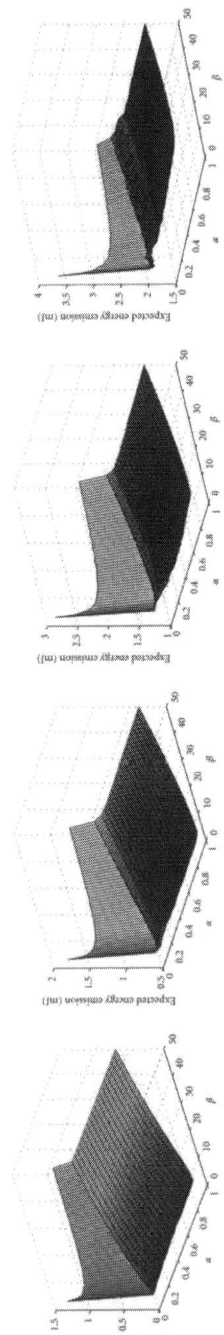

Figure 8. 3D α and β combinations for different Scenarios (802. 15.4).T1T2T3T4

Table 2. Quantitative comparison of expected energy emission for the updating phase: IEEE 802.15.4.

Scenario	T1	T2	T3	T4
Fixed (mJ)	1.3496	1.8223	2.8932	3.6682
PDR-based (mJ)	0.2735	0.5865	1.1863	1.9430
Reduction	−80%	−68%	−59%	−47%
PDR-based (optimal) (mJ)	0.1428	0.5621	1.11572	1.9118
Optimal α	0.15	0.5	0.35	0.35
Optimal β	2%	10%	14%	28%

(I) Data Rate

IEEE 802.11 offers multiple data rates. Based on different channel conditions, different modulation and coding schemes can be used to achieve the highest throughput. The main question is whether the results in Section 6.1.2 are valid for other data rates or not. Moreover, from the energy emission perspective, the highest throughput data rate may not be the one that saves the most energy. We post processed the measurement traces shown in Figure2(b) to see the impact of α and β and display it in Figure 9(a).

We can see that for the three experiments with lower data rates (2, 5.5, 6 Mbps), the effect is the same as in Figure6(a). However, for the high rates (11, 54 Mbps), due to the lower PDR at almost all the transmission power levels, the updating phase cannot do much to reduce energy emission and 15 dBm is optimal. Therefore, α does not have much impact on the performance. Since these experiments were done in the same location, we can see that 5.5 Mbps is the most power saving data rate. However, if we change to another location with another distance, another data rate may be optimal, which suggests that not only the transmission power level, but also the data rate can be selected to save power.

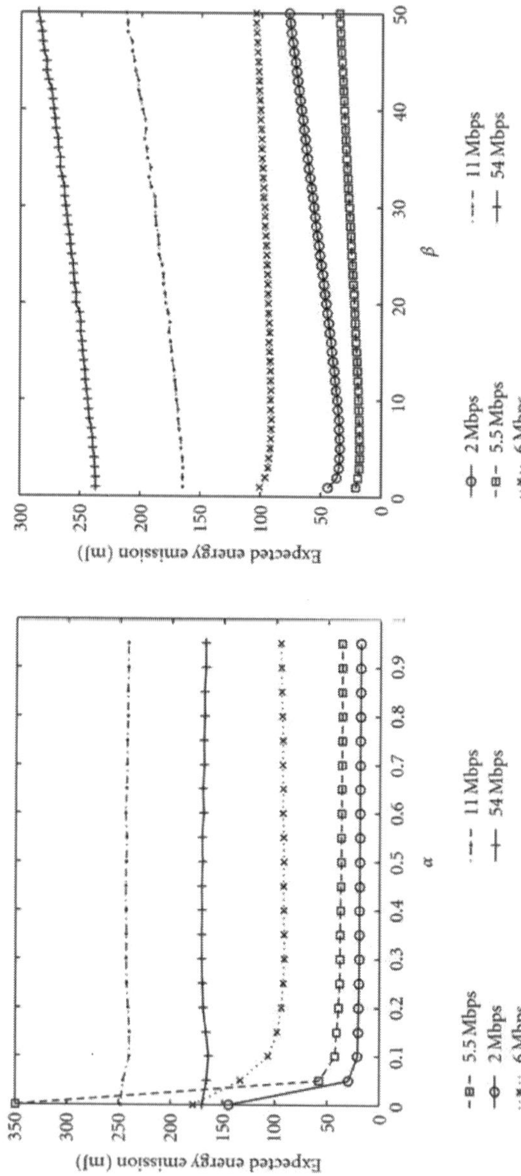

Figure 9. The best α **and** β **for IEEE 802.** 11 with difference data rates.best α best β

Concerning β, low data rates need more probe packets to update the PDR information. For the higher data rates in this experiment, most other power levels will directly result in packet drops. Therefore, probing more on other power levels increases the expected energy emission. In other scenarios, this may not be the case.

(ii) Packet Size

It is also important to know the performance with different packet-sizes. In order to experiment with different packet-sizes over a similar link, we used 802.15.4 and changed the packet sending method. For each packet-size, we sent 30000 packets in 200 batches. We splitted each of the three different packet-size experiments into 10 equalled sized parts and we interleaved the parts of the experiments in a round robin fashion. In this way, the three packet-sizes experiment are experiencing very similar channel conditions. By interleaving, we mitigate unfairness caused by slowly changing link conditions.

We again calculated the best α and β using our post processing method and got the results in Figure 10. For the best β, the conclusion is similar as previously. However, it is interesting to see the result for the best α. When $\alpha = 0$, the energy emission is still very high. However, a larger α will have less energy emission. To deliver the same amount of information, a larger packet-size emits less energy in this scenario.

(iii) Mobility

Since all the scenarios are stationary, we still need to investigate the performance in a mobile environment. Instead of moving the devices around, which is time consuming and difficult to replicate between experiments, we emulated a mobile channel by stitching together the traces from scenario T1, T2, and T3. The way we generate mobility is not ideal, but it still gives us some insights. It shows how channel variance can affect the selection of α and β, which is our interest.

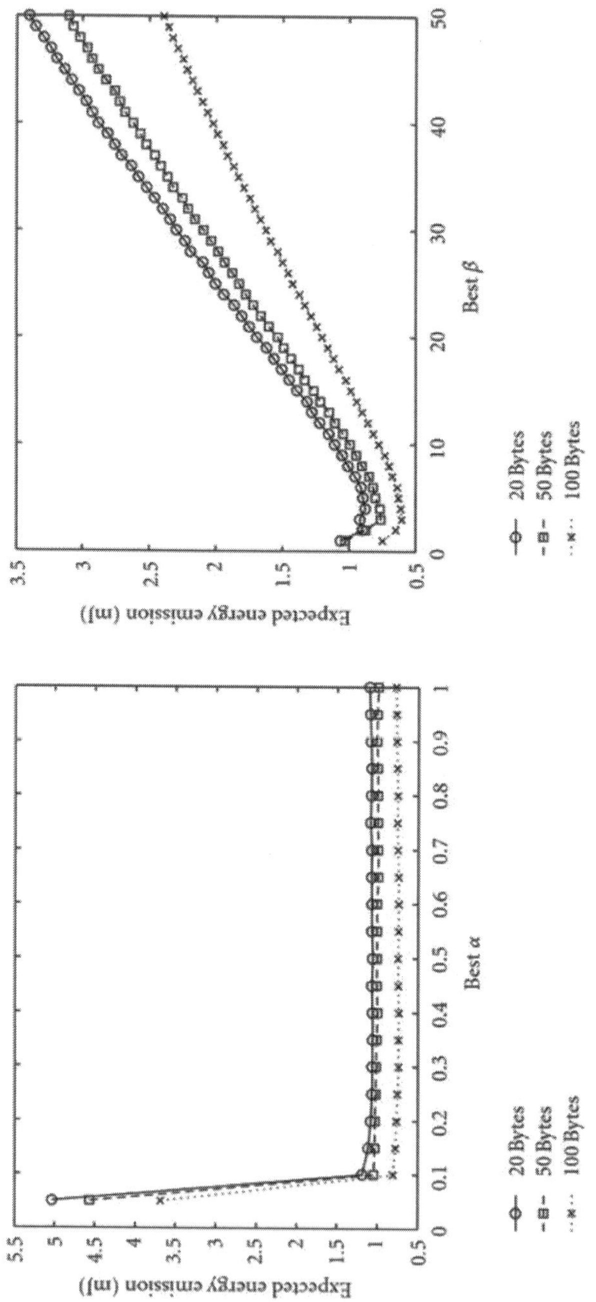

Figure 10. The best α and β for IEEE 802. 15.4 with difference packet-sizes.best α best β

The measurement results were segmented into different numbers of batches based on different mobility level assumptions. The mobile channel went from good (T1), to medium (T2), to bad quality (T3). To emulate different amount of mobility, we stayed with the same scenario data for different amount of time before changing to the next scenario trace. We used four different PDR change speeds, which we measured in number of batches. A mobility scenario which changes trace every 10 batches, that is, Mobility-10, is more mobile than one which changes only every 100 batches, that is, Mobility-100.

We plot the performance of different α for different mobility levels in Figure 11. We can see that for low mobility levels, for example, Mobility-100, the best α is not big. When the mobility level increases, the best α also increase. This conclusion does not hold for the situation with Mobility-10, in which the mobility is quite high. It is interesting to note that despite different mobility levels, the expected minimum energy emission remains similar. Less mobility will decrease the minimum expected energy emission, but not significant.

We also plotted all the combinations of the α and β for different mobility levels with 3D graphs in Figure 12. The results are similar to the previous 3D results in Figure 8. However, it is obvious that with higher mobility, the best αvalue gradually increases, except for Figure 12(a). Perhaps somewhat surprising, different mobility levels do not cause too much impact on the optimal β value. Furthermore, the minimum expected energy emission is similar for all levels of mobility. Hence, mobility does not affect α and β very much.

(iv) Variance over Time

To study the change of the optimal α value over time, we segmented the experiments into 5 different pieces in the time domain and used the same method as previously to select the best α for each piece. The results for both IEEE 802.11 and 802.15.4 are shown in Figure 13. We found that the optimal selection for α changes over time for all links. However, the difference in energy emission between the different α

values is not significant, given the results in Figure 6(a). Therefore, in practice, one should just stick to one fixed α as long as $\alpha \geq 0.1$. For IEEE 802.15.4, the variance of the optimal α is larger over time, but still the energy emission difference is small as shown in Figure 7(a).

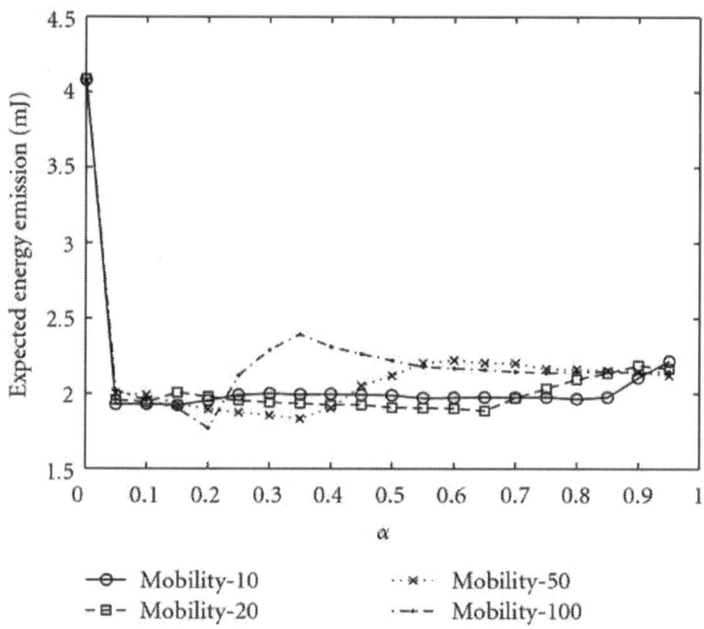

Figure 11. The best alpha value with different levels of mobility.

6.1.4. Discussion

Although all experiments were done with only changing the transmission power levels, our PDR-based method can also be applied to multi-factor environments. For example, we can keep records of three different data rates (or packet-sizes) but only five different power levels instead of all 15. The selection would then involve finding the most optimal rate (packet-size) and power level combination. It is also obvious to see that if proper α and β values are selected, the energy emission can be reduced in most scenarios.

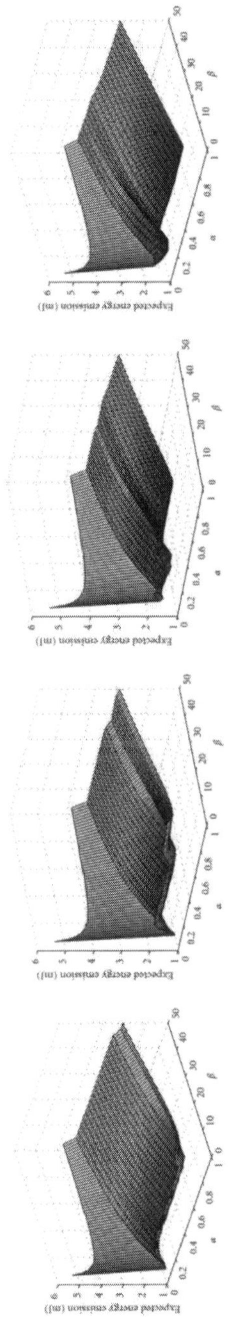

Figure 12. α **and** β **combinations for different mobility levels.** Mobility-10Mobility-20Mobility-50Mobility-100

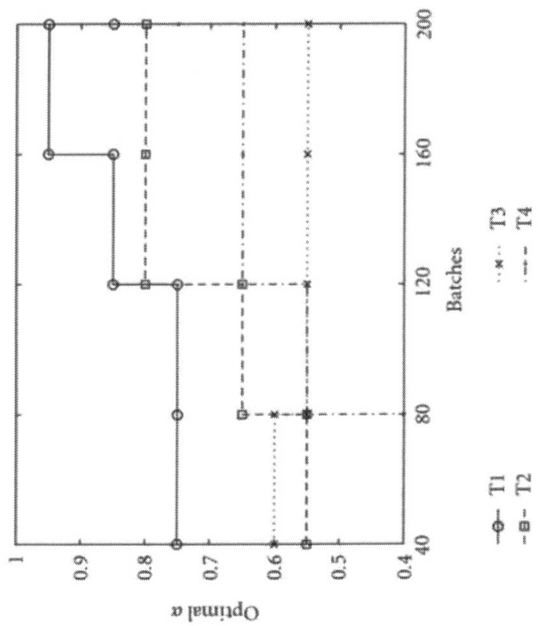

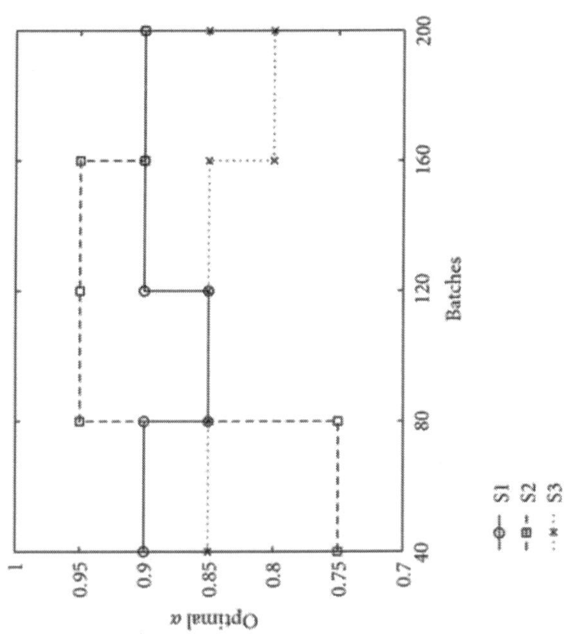

Figure 13 The best α with time for IEEE 802. 11 and IEEE 802.15.4.IEEE 802.11IEEE 802.15.4

6.2. Energy Consumption Reduction

Since energy saving is more important for IEEE 802.15.4 devices, we also need to look at reducing the energy consumption by means of selecting the transmission power level. To do this, we used the trace files of the IEEE 802.15.4 radios from scenario T1, T2, and T4. However, we replaced the energy model and used $P = 35 \cdot P_{RF} + 30$instead. Then we did the same experiments as in the previous section. We show the results in Figure 14 and we can see that we got similar conclusions with respect to α and β as the results in Figure 7. The main difference is that the expected energy consumption is much larger than the expected energy emission due to the fact that most consumed energy is not emitted. Due to the similarity of the results, we do not present more results for the energy consumption calculation for other situations. Instead, to better demonstrate the effectiveness of our mechanism, we compare it with a typical signal strength-based algorithm proposed in [15]. We used their *aggressive method* in our comparison, since it was claimed by the authors to have better performance. We briefly introduce their mechanism as follows.

The mechanism uses the received packets' signal strength to adapt the transmission power level. This received signal strength is denoted as R . An EWMA method (see (5)) is used to smooth R with $\alpha = 0.8$ and then denoted as $\hat{R}$. Every time a packet is received, $\hat{R}$ is updated and then one of the following is done:

$$\begin{aligned}
&\text{if } \hat{R} < T_L, \quad \text{double the transmit power;}\\
&\text{if } T_L \le \hat{R} \le T_H, \quad \text{keep the same transmit power;}\\
&\text{if } \hat{R} > T_H, \quad \text{reduce the transmit power by a constant;}
\end{aligned} \tag{6}$$

where $T_L = -85$ dBm and $T_H = -80$ dBm for our wireless chip CC2420 according to [15].

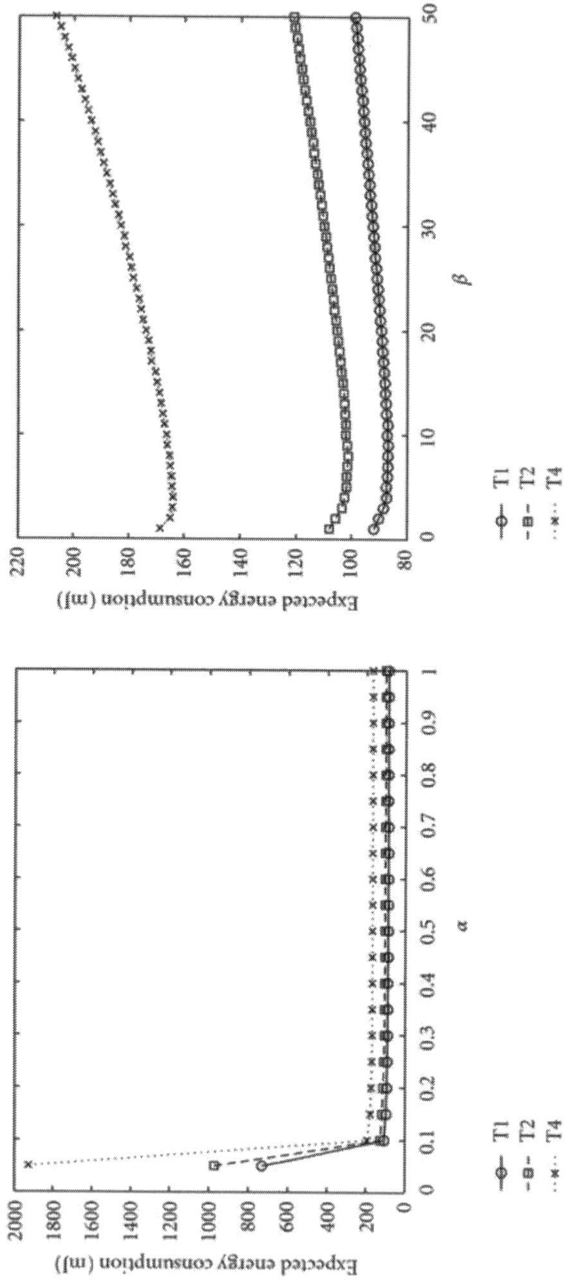

Figure 14. The best α and β for IEEE 802. 15.4 energy consumption, 20 Bytes.best α best β

During the implementation, we found that there is a logical error in the mechanism. If a packet is lost, no new signal strength value will be read and $\hat{R}$ will not be updated. If, at the same time, $\hat{R} > T_L$, the transmission power will not be increased and this can lead to a deadlock. This situation could be quite possible, especially in mobile scenarios. Therefore, we adapted their method by letting a lost packet have an assumed received signal strength of −95 dBm, the lowest receivable signal strength for CC2420.

Furthermore, the authors do not specify the constant used when decreasing the transmission power level. In our implementation, we decrease the power by one level (corresponding to a decrease between 1 dBm and 3 dBm depending on the levels) when $\hat{R} > T_H$.

To compare the two mechanisms, we used scenario T1, T2, and T4. The same trace data was used, since we also recorded the received signal strength. The results are shown in Table 3. We can see that the PDR-based method outperformed both the Fixed method and the signal strength method. Our PDR-based method always saves energy compared to the Fixed method, meanwhile, the signal strength-based method sometimes performs worse. This is because it only focuses on signal strength and ignores packet loss and at the same time, packet loss increases so much that the increased number of retransmissions cancels out the lower power consumption of using a lower transmission power level.

Table 3. Quantitative comparison for the energy consumption of IEEE 802.15.4.

Scenario	T1	T2	T4
Fixed (mJ)	154.0	154.8	169.9
Signal strength-based (mJ)	121.6	154.8	169.9
PDR-based (mJ)	79.9	130.2	164.1

The reason why the PDR-based method performs better is its direct use of the PDR to power level correlation. The signal strength-based method uses fixed thresholds for all the scenarios, which may work in some cases, but not in others, such as T1 and T2. The correlation between received signal strength and packet loss is simply too weak.

To show the reason behind the good performance of the PDR-based method, we show in Figure 15 the transmission power level distribution for both methods in scenario T1 and T2 as examples. Also compare these results with the PDRs for these levels shown in Figure 3(a) and the energy consumption shown in Figure 3(e). We can clearly see that the signal strength method always use the maximum power level in scenario T2, which is the same as the Fixed method. The same happens in scenario T4 (not shown). Only in scenario T1, it shows some "smarter" selection. For the PDR-based method, we can clearly see that the transmission power level used for most packets are close to the optimal level 7 (see Figure 3(e)).

6.3. Trade-Off between Energy Emission and Consumption

In some cases, a trade-off between minimizing the energy emission and the energy consumption may be sought. To allow for this, we introduce a tunable parameter ω as follows.

When comparing (2) with (3) or (4) and given that a multiplicative scalar does not affect the end result, we may capture all optimization functions with the following equation:

$$P = P_{\mathrm{RF}} + \omega, \qquad (7)$$

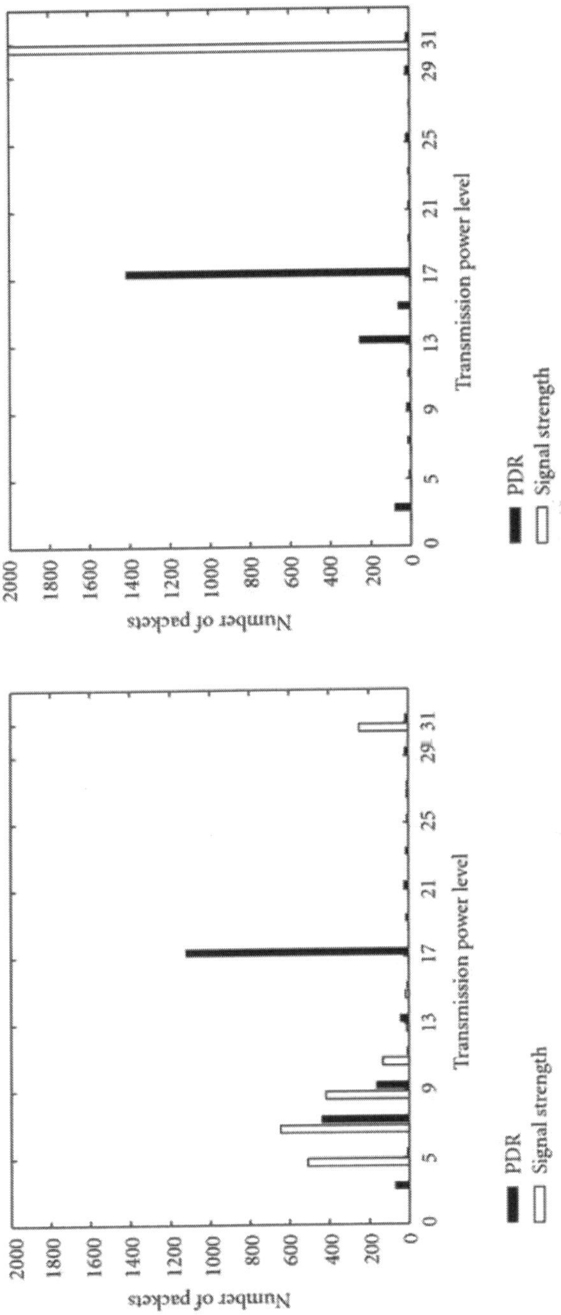

Figure 15. The transmission power levels distribution for the proposed and signal strength-based algorithms in the two scenarios. T1T2

where $\omega \in [0, 1400/10]$ for IEEE 802.11 or $\omega \in [0, 30/35]$ for IEEE 802.15.4. If we set $\omega = 0$, we get (2) and will optimize for energy emission. If we set $\omega = 1400/10$ for IEEE 802.11 (or $\omega = 30/35$ for IEEE 802.15.4), we optimize for energy consumption. However, it is also possible to set ω to a value in between and that would mean that we get a trade-off between energy emission and energy consumption.

To demonstrate how this tuning works, we used the T1 scenario as an example and calculated the expected energy consumption and emission with different ω values. The result is plotted in Figure 16. We can see how different ωvalues affect the energy emission and energy consumption for this scenario. By making ω smaller, we reduce the energy emission, while the energy consumption increases and by making ω larger, we reduce the energy consumption, while the energy emission increases. Hence, we can use ω as a tuning parameter.

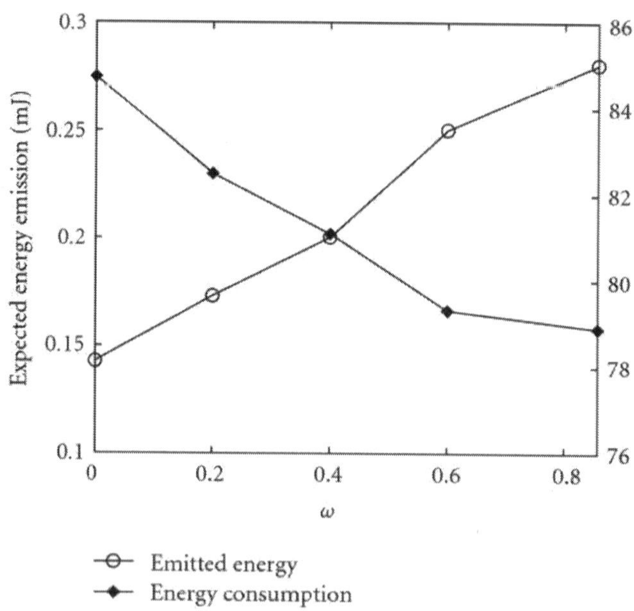

Figure 16. Trade off between minimum energy consumption and energy emission.

We can also use ω to control the expected delay. A larger ω will penalize a lost packet more and this causes a higher power level to be chosen. The effect is fewer packet losses, fewer retransmissions, and thereby less delay.

7. CONCLUSIONS

Energy emission and consumption reduction are key challenges for wireless networks. In this paper, we proposed to select the appropriate transmission power level using PDR information in order to reduce the energy emission and/or consumption. We proposed to use the EWMA method to update the PDR and transmission power level correlation and use this to adapt the transmission power. We proposed four initialization phase methods to get a good transmission power level to start with. Furthermore, we investigated the optimal parameters for this correlation in order to achieve minimum energy emission or consumption. Different impacting factors were also analyzed. We carried out measurements in two types of test-beds and showed that a significant amount of energy can be saved for the transmitter in typical scenarios. We also compared our mechanism with a signal strength-based mechanism and showed improved energy savings. Finally, we demonstrated that our mechanism can be tuned to achieve a balance between the minimum energy consumption and emission, which enables the user to adaptively set the desired target.

REFERENCES

1. Li Q, Aslam J, Rus D: Online power-aware routing in wireless ad-hoc networks. *Proceedings of the 7th Annual International Conference on Mobile Computing and Networking (Mobicom '01), July 2001* 97-107.

2. Choi J-M, Ko Y-B, Kim J-H: Enhanced power saving scheme for IEEE 802.11 DCF based wireless networks. *Personal Wireless Communications, 2003, Lecture Notes in Computer Science* 2775: 835-840.

3. Shrivastava V, Agrawal D, Mishra A, Banerjee S, Nadeem T: Understanding the limitations of transmit power control for indoor WLANs. *Proceedings of the 7th ACM SIGCOMM Internet Measurement Conference (IMC '07), October 2007* 351-364.

4. Lal D, Manjeshwar A, Herrmann F, Uysal-Biyikoglu E, Keshavarzian A: Measurement and characterization of link quality metrics in energy constrained wireless sensor networks. *Proceedings of IEEE Global Telecommunications Conference (GLOBECOM '03), December 2003, San Francisco, Calif, USA* 1: 446-452.

5. Kim J, Huh J: Link adaptation strategy on transmission rate and power control in IEEE 802.11 WLANs. *Proceedings of the 64th IEEE Vehicular Technology Conference (VTC '06), September 2006* 2053 2057.

6. Wattenhofer R, Li L, Bahl P, Wang Y: Distributed topology control for power efficient operation in multihop wireless ad hoc networks. *Proceedings of the 20th Annual Joint Conference of the IEEE Computer and Communications Societies (INFOCOM '01), 2001, Anchorage, Alaska, USA* 3: 1388-1397.

7. Lin S, Zhang J, Zhou G, Gu L, Stankovic JA, He T: ATPC: adaptive transmission power control for wireless sensor networks. *Proceedings of the 4th International Conference on Embedded Networked Sensor Systems (SenSys '06), 2006, Boulder, Colo, USA* 223-236.

8. Hackmann G, Chipara O, Lu C: Robust topology control for indoor wireless sensor networks. *Proceedings of the 6th ACM Conference*

on Embedded Network Sensor Systems (SenSys '08), 2008, Raleigh, NC, USA

9. Correia LHA, Macedo DF, Silva DAC, Dos Santos AL, Loureiro AAF, Nogueira JMS: Transmission power control in MAC protocols for wireless sensor networks. *Proceedings of the 8th ACM Symposium on Modeling, Analysis and Simulation of Wireless and Mobile Systems, October 2005, Montreal, Canada* 282-289.

10. Son D, Krishnamachari B, Heidemann J: Experimental study of the effects of transmission power control and blacklisting in wireless sensor networks. *Proceedings of the 1st Annual IEEE Communications Society Conference on Sensor and Ad Hoc Communications and Networks (SECON '04), 2004* 289-298.

11. Monks JP, Bharghavan V, Hwu WW: A power controlled multiple access protocol for wireless packet networks. *Proceedings of the 20th Annual Joint Conference on the IEEE Computer and Communications Societies (INFOCOM '01), 2001, Anchorage, Alaska, USA* 1: 219-228.

12. Jung E-S, Vaidya NH: A power control MAC protocol for ad hoc networks. *Wireless Networks* 2005, 11(1-2):55-66.

13. Agarwal S, Katz RH, Krishnamurthy SV, Dao SK: Distributed power control in ad-hoc wireless networks. *Proceedings of the 12th International Symposium on Personal, Indoor and Mobile Radio Communications (PIMRC '01), October 2001, San Diego, Calif, USA* F59-F66.

14. Kubisch M, Karl H, Wolisz A, Zhong L, Rabaey J: Distributed algorithms for transmission power control in wireless sensor networks. *Proceedings of IEEE Wireless Communications and Networking (WCNC '03), 2003, New Orleans, La, USA* 1:

15. Xiao S, Dhamdhere A, Sivaraman V, Burdett A: Transmission power control in body area sensor networks for healthcare monitoring.

IEEE Journal on Selected Areas in Communications 2009, 27(1):37-48.

16. Chipcon : CC2240: 2.4GHz IEEE 802.15.4 /ZigBee-ready RF Transceiver. http://www.chipcon.com

17. Liaskovitis P, Schurgers C: Energy consumption of multi-hop wireless networks under throughput constraints and range scaling. *ACM SIGMOBILE Mobile Computing and Communications Review* 2009, 13: 1-13.

18. Williamson C: Internet traffic measurement. *IEEE Internet Computing* 2001, 5(6):70-74.

Index

A

Active control, 2, 7
Adaptive waveform matching extension, 70
Artificial neural networks, 100

C

Cumulants, 100
Current Harmonic Perturbations, 62

E

Eigenvalue Analysis, 43
End effect, 70, 72, 79

F

flicker detection, 86
Flow battery, 176, 190, 200
Fluke harmonic experiment, 91
Fresnel Lens, 219, 220, 231, 235, 236

H

Harmonic analysis, 14
Higher-order statistics, 100, 102, 105, 120, 121
hydro power plant, 176, 184, 187, 193, 194, 197, 201, 205, 207, 216

I

Improved Hilbert-Huang transform, 70
Integer harmonic detection, 81

L

Linear Programming, 124

M

Microgrid, 70
microgrids, 1, 2, 12, 13, 14, 21, 24, 25
Mode mixing, 74
Multi-Machine Power System, 28

N

Nyquist Diagram, 28

P

Passive control, 2, 4
Power quality, 2, 3, 4, 16, 18, 23, 52, 70, 100, 118, 119, 172
Power system balance, 176
Power Systems, 27, 47, 48, 51, 119, 124, 145, 146, 175, 213
Probabilistic Analysis, 124, 141

Q

Quality Assessment, 124, 126, 145, 146

R

Regression algorithms, 100, 116
Reliability, 123, 124, 129, 131, 145, 146
Robust Control, 28, 29, 48

S

Simultaneous reactive power injection, 148
Smart grid, 2, 100
Solar Concentrator, 219, 220, 231, 235, 239
Solar Energy, 220, 238, 239
Spectral kurtosis, 100
Steamgeneration, 220

T

Tow Area Four Machines Test System, 36

U

Unified power quality conditioner, 147, 148

V

Voltage Fluctuations, 53
Voltage sag detection, 81

W

Wind farms, 176
Wind Penetrations, 28, 43
Wind power generation, 176
wireless device, 247

Z

Zero active power injection compensation region, 148